Understanding Innovation

Series Editors

Christoph Meinel
Larry Leifer

For further volumes:
http://www.springer.com/series/8802

Hasso Plattner • Christoph Meinel • Larry Leifer
Editors

Design Thinking Research

Building Innovation Eco-Systems

Springer

Editors
Hasso Plattner
Christoph Meinel
Hasso-Plattner-Institute
Campus Griebnitzsee
Potsdam
Germany

Larry Leifer
Center for Design Research (CDR)
Stanford University
Stanford
CA
USA

ISBN 978-3-319-03290-0 ISBN 978-3-319-01303-9 (eBook)
DOI 10.1007/978-3-319-01303-9
Springer Cham Heidelberg New York Dordrecht London

Printed on acid-free paper

Springer is part of Springer Science+Business Media (www.springer.com)

Preface

In this volume of Design Thinking Research, 'Building Innovation Ecosystems' // 'Building Multilateral Innovation Ecosystems', the results are combined of the Design Thinking joint venture of the Hasso Plattner Institute of Design located at Stanford University in Palo Alto, California, USA, and the Hasso Plattner Institute for IT Systems Engineering located in Potsdam, Germany.

Design Thinking is everywhere. Every approach in normal life to find solutions for everyday problems involves thinking. Synthesized thinking with specialized procedures conducted in the best practical manner reflects the results in this book.

Nevertheless, the goal of this book is not only to show what's possible in the field of Design Thinking but also to show how and why it works so well. Therefore, this book collects multiple points of view on multiple problems, companies and technologies as well as the solution for wicked and real-world problems, views of the management position on Design Thinking and the advantage it can achieve for engineering software projects or even building prototypes in manufacturing.

Design Thinking is no unyielding concept. It's a way of interacting within your environment, of interacting with your colleagues and friends and, of course, it's your own physical and mental interaction in a differentiated and fast-paced world; in short, it's an ecosystem, where innovations are nurtured to growth. The methods, techniques and considerations of the specialties of Design Thinking help to organize the world. This means to see what's happening and to react accordingly in the world and to construct a way out of complex problems, which usually have not only one simple solution but often a multifaceted, correspondingly complex one.

This book not only formulates many ways out of complex problems and situations but also shows, either explicitly or implicitly, what can be done with an open mind, a clear vision and the strength to envision problems as challenges and to foster ways of solution-finding.

That's what this book is about and that's what the Hasso Plattner Institutes are about: a clear vision with energy and strength to find solutions where others give up.

Palo Alto/Potsdam
Winter 2012/2013

Prof. Dr. h.c. Hasso Plattner

Contents

Part I All Design Activity Is Ultimately Social in Nature

Introduction 3
Christoph Meinel and Larry Leifer

Student Teams in Search of Design Thinking 11
Shelley Goldman, Zandile Kabayadondo, Adam Royalty, Maureen P. Carroll, and Bernard Roth

Team Cognition and Reframing Behavior: The Impact of Team Cognition on Problem Reframing, Team Dynamics and Design Performance 35
Greg Kress and Joel Sadler

Early and Repeated Exposure to Examples Improves Creative Work 49
Chinmay Kulkarni, Steven P. Dow, and Scott R Klemmer

Part II Design Thinkers Must Preserve Ambiguity

Impact and Sustainability of Creative Capacity Building: The Cognitive, Behavioral, and Neural Correlates of Increasing Creative Capacity 65
Grace Hawthorne, Eve Marie Quintin, Manish Saggar, Nick Bott, Eliza Keinitz, Ning Liu, Yin Hsuan Chien, Daniel Hong, Adam Royalty, and Allan L. Reiss

Acting with Creative Confidence: Developing a Creative Agency Assessment Tool 79
Adam Royalty, Lindsay Noelle Oishi, and Bernard Roth

How Design Thinking Tools Help To Solve Wicked Problems 97
Julia von Thienen, Christoph Meinel, and Claudia Nicolai

Part III All Design Is Re-design

How Prototyping Helps to Solve Wicked Problems 105
Birgit Jobst and Christoph Meinel

Creative Collaboration in Real World Settings 115
Matthias Wenzel, Lutz Gericke, Raja Gumienny, and Christoph Meinel

User-Centered Innovation for the Design and Development of Complex Products and Systems 135
Lauren Aquino Shluzas, Martin Steinert, and Riitta Katila

Part IV Make Ideas Tangible

Connecting Designing and Engineering Activities 153
Thomas Beyhl, Gregor Berg, and Holger Giese

A Research Plan for the Integration of Design Thinking with Large Scale Software Development Projects 183
Thomas Kowark, Franziska Häger, Ralf Gehrer, and Jens Krüger

Sharing Knowledge Through Tangible Models: Designing Kickoff Workshops for Agile Software Development Projects 203
Markus Guentert, Alexander Luebbe, and Mathias Weske

How to Compare Performance in Program Design Activities: Towards an Empirical Evaluation of CoExist 219
Bastian Steinert and Robert Hirschfeld

Design Thinking: Expectations from a Management Perspective 239
Holger Rhinow and Christoph Meinel

Part I
All Design Activity Is Ultimately Social in Nature

Introduction

Christoph Meinel and Larry Leifer

1 Why Do We Need Innovation-Eco-systems?

We have seen over the years that many individuals appreciate the power of engineering design thinking. At the same time we witness an almost unfathomable skepticism about the ability of established organizations (corporate, academic, and government) to really adopt the paradigm. Some argue that the paradigm only works in the world of "start-ups".

Motivated by this dilemma, we ask how the fundamental rules of engineering design thinking might be translated into design requirements for building precision innovation eco-systems.

1.1 Rules of Design Thinking

We now have evidence in support of several design thinking activities that have long been considered important, but until this time we have not had an explanation or understanding of their value. Of these, the over-arching truth lies in the fact that every physical product delivers a service; that every service is manifest through physical products. Our research suggests that four "rules of design thinking" are particularly relevant. The challenge of this section is to translate these rules into innovation eco-system design requirements.

C. Meinel (✉)
Internet Technologies and -Systems, Hasso Plattner Institute for Software Systems Engineering, Campus Griebnitzsee, Postdam 14482, Germany
e-mail: christoph.meinel@hpi.uni-potsdam.de

L. Leifer
Stanford Center for Design Research, Stanford University, Panama Mall 424, Stanford 94305-2232, CA, USA
e-mail: leifer@cdr.stanford.edu

H. Plattner et al. (eds.), *Design Thinking Research*, Understanding Innovation,
DOI 10.1007/978-3-319-01303-9_1,

I. The Human Rule: All Design Activity is Ultimately Social in Nature. Never Go Hunting Alone.

There are studies that substantiate the assertion that successful innovation through design thinking activities will always bring us back to the "human-centric point of view". This is the imperative to solve technical problems in ways that satisfy human needs and acknowledge the human element in all technologies and organizations.

Design Requirements for the Human Rule

Place people at the center of all things. Cover the walls with images of your people. Celebrate their success and failures. Build environments that celebrate the "human scale". Forgo monuments. Create an atmosphere of empathy-in-action; for yourself and others.[1]

II. The Ambiguity Rule: Design Thinkers Must Preserve Ambiguity. Never Go Home Empty Handed.

There is no chance for "chance discovery" if the box is closed tightly, the constraints enumerated excessively, and the fear of failure always at hand. Innovation demands experimentation at the limits of our knowledge, at the limits of our ability to control events, and with freedom to see things differently.

Design Requirements for the Ambiguity Rule

Keep track of assumptions; place them boldly in your design space. For every constraint you are coping with, list a competing opportunity. Check your thinking: are you're looking for the global fix, or, are you keenly aware that most everything in design and business is context dependent. Take time to define the problem and solutions space context.

III. The Re-Design Rule: All Design Is Re-Design. Take the Big Idea Home. It Has Been Done Before.

The human needs that we seek to satisfy have been with us for thousands of years. Through time and evolution there have been many successful solutions to these problems. Because technology and social circumstances change constantly, it is imperative to understand how these needs have been addressed in the past. Then we can apply "foresight tools and methods" to better estimate the social and technical conditions we will encounter 5, 10, 20 years in the future.

Design Requirements for the Re-Design Rule

Hunting is tough. Taking it home is tougher. Nothing beats a prepared mind; be sure your team is well informed about the history of organizational change and context. How did others effect change? How did they circumnavigate the skeptics? Take advantage of foresight thinking and tools.[2]

[1] Kress, 2012: http://purl.stanford.edu/hm975hz5458

[2] Cockayne, 2013: http://foresight.stanford.edu/index.html

IV. The Tangible Rule: Make Ideas Tangible. Facilitate Human Communication.

Curiously, this is one of our most recent findings. While conceptual prototyping has been a central activity in design thinking during the entire period of our research, it is only in the past few years that we have come to realize that "prototypes are communication media". Seen as media, we now have insights regarding their bandwidth, granularity, time constants, and context dependencies. The "make it tangible" rule is one of the first major findings of the design thinking research program documented in this book.

Design Requirements for the Tangible Rule

There are more great ideas out there in the world than inside our heads. Put differently, searching in the world tangibly is a great way to get new ideas, unplanned associations, un-dreamed metaphors, serendipity squared. Show me, don't tell me.[3]

We have summarized and in some cases paraphrased the design requirements in the following table. Take the framework and apply it to your project, your organization, and your team. This is not a tool of physics. Everything about it is context dependent. Define your context.

1.2 Innovation Eco-system Design Requirements

Requirement	Context	Metric	Rationale
The Human Rule All design activity is ultimately social in nature. Never go hunting alone.	Every human eco-system is unique. Every business scenarios is unique. Take time to observe and document your context design context. Where are you hunting?	Count the people in your framework. Count your team's linkages. Cover the walls with images of your team, the users, the competition. Count their success and failures. Count the experiments tried. Share the learnings.	Nurture an atmosphere of "empathy-in-action".[4] No other single variable has been documented to more concretely contribute to team performance.
The Ambiguity Rule Design Thinkers must preserve ambiguity. Never go home empty handed.	Check your thinking; are you looking for the global fix, or, are you keenly aware that most everything in design and business is context dependent. What do you hunt for?	Count your assumptions, opportunities, and constraints. How many ways can there be defined? Don't let others shut you down by stealing your ambiguity.	Ambiguity is the mutable stuff that allows for creativity. It is the foundation material of "creative self-efficacy."[5]

(continued)

[3] Edelman, 2012: http://purl.stanford.edu/ps394dy6131

[4] Kress, 2012: http://purl.stanford.edu/hm975hz5458

[5] Albert Bandura: http://en.wikipedia.org/wiki/Albert_Bandura

Requirement	Context	Metric	Rationale
The Re-design Rule All design is re-design. Take the big idea home. It has been done before.	Most human needs have been satisfied before. Know the context of past solutions. How do they map to the foreseeable future. Understand past hunters and the hunted.	Count the number of ways this need has been satisfied in the past. Enumerate the pros and cons. Position your team to absolutely nail just one of the cons without losing the pros.	Foresight innovations is unexpectedly all about understanding the past to be ready for the future. Never leave home without it.[6]
The Tangible Rule Make ideas tangible. Facilitate human communication.	There are more great ideas out there in the world than inside our heads. Searching tangibly is a great way to get ideas, associations, metaphors, serendipity squared.	Count the tangible encounters. Amplify the energy in the room. Keep evidence of past successes and failures at hand.	Show me, don't tell me.[7]

2 The HPI Design Thinking Research Program

Started in 2008, the HPI Design Thinking Research program is financed and supported by the Hasso Plattner Foundation.

2.1 *Program Vision*

The HPI-Stanford Design Thinking Research Program engages multidisciplinary research teams to scientifically investigate the phenomena of innovation in all its holistic dimensions. In particular, researchers are encouraged to develop ambitious, long-term explorations related to the innovation methods of design thinking in its technical, business, and human aspects. The HPI-Stanford Design Thinking Research Program strives to apply rigorous academic methods to understand the scientific basis for how and why the innovation method of design thinking work and fail. Researchers in the Program study e.g. the complex interaction between members of multi-disciplinary teams challenged to deliver design innovations. An important feature of the domain is the need for creative collaboration across spatial, temporal, and cultural boundaries. In the context of disciplinary diversity, how do design thinking methods mesh with traditional engineering and management approaches, specifically, why does the structure of successful design teams

[6] Cockayne, 2013: http://foresight.stanford.edu/index.html

[7] Edelman, 2012: http://purl.stanford.edu/ps394dy6131

differ substantially from traditional corporate structures? The overall goal of the program is to discover metrics that determine the success of challenges approaches with design thinking methods. A special program interest is to explore the use of design thinking in the field of Information technology and IT systems engineering.

2.2 Program Priorities

The focus of the Design Thinking Research Program is on the collaboration between researchers of the Stanford University and the Hasso Plattner Institute, Potsdam, Germany. The funding favors projects that set new research priorities for this emergent knowledge domain. In this context we believe that the field studies in real business environment are important to assess the impact and/or needed transformations of Design Thinking in organizations. Selection is also based on intellectual merit and evidence of open collaboration.

The following guiding research questions are of special interest:

- What are people really thinking and doing when they are engaged in creative design innovation? How can new frameworks, tools, systems, and methods augment, capture, and reuse successful practices?
- What is the impact of Design Thinking on human, business, and technology performance? How do the tools, systems, and methods really work to create the right innovation at the right the time? How do they fail?

2.3 Road Map Through This Book

The organization of the material presented in this book follows the four rules: "The Human Rule", "The Ambiguity Rule", "The Re-design Rule" and "The Tangible Rule".

2.4 Part I: All Design Activity Is Ultimately Social in Nature

The article "Student Teams In Search Of Design Thinking" is by Shelley Goldman, Zandile Kabayadondo, Adam Royalty, Maureen P. Carroll, and Bernard Roth. It's about students who learn the design thinking process. It investigates how design thinking reflects on their work and explores and analyzes their communication as they interact.

"Team Cognition and Reframing Behavior". The Impact of Team Cognition on Problem Reframing, "Team Dynamics and Design Performance" is about the mental framework while doing design thinking and how this frame building works, as well as how a reframing takes place when new facts and new solutions are faced. It's written by Greg Kress and Joel Sadler.

Chinmay Kulkarni, Steven P. Dow and Scott R. Klemmer explore in "Early and Repeated Exposure to Examples Improves Creative Work" online creativity experiment and the outcome of the effect of sample timing on this creativity. The between-subject experiment deals with repeated exposure to the experimental setting and its iterative experimental results.

Examination of the issue of whether creativity can be acquired or learned by an individual over time and how this relates to cognition, behavior and the brain is the last article in this part. The focus is on cognitive, behavioral and neural processes that may contribute to creative capacities. It's entitled "Impact and Sustainability of Creative Capacity Building", by authors Grace Hawthorne, Eve-Marie Quintin, Manish Saggar, Nick Bott, Eliza Keinitz, Ning Liu, Yin-Hsuan Chien, Daniel Hong, Adam Royalty, and Allan Reiss.

2.5 Part II: Design Thinkers Must Preserve Ambiguity

The authors of the article "Acting with Creative Confidence: Developing a Creative Agency Assessment Tool", written by Adam Royalty, Lindsay Oishi, and Bernard Roth, deal with the problem of enabling new technologies in learning models aimed at developing creativity and innovation, without checking their outcome. This article uses quality and quantity test results on behalf of new criteria models.

Captioned with the title "How Design Thinking Tools Help to Solve Wicked Problems" the article speaks about a psychological way of thinking and a perspective on solving such problems. Julia von Thienen, Christoph Meinel and Claudia Nicolai move in a field of design thinking tools where by their very nature these tools offer the necessary support to answer wicked challenges.

The next article in this part continues the thoughts of the preceding. This text is entitled "How Prototyping Helps to Solve Wicked Problems." The authors are Birgit Jobst and Christoph Meinel. They explore how wicked problem solutions can be figured out by improved prototyping skills and the relations between three major factors in this process: prototyping skills, problem solving competence and the issue of self-efficacy.

2.6 Part III: All Design Is Re-Design

The article "Creative Collaboration in Real World Settings" written by Matthias Wenzel, Lutz Gericke, Raja Gumienny, and Christoph Meinel is about the Tele-Board system. This system was designed and implemented especially for design thinking teams, but is of equal interest to many different companies who use it worldwide. This article presents the results of a study based on Tele-Board's use by a global company and shows how it could meet the company's requirements.

In addition, showing Tele-Board's multi-faceted implementation, areas of further development are also discussed.

In "User-Centered Innovation For The Design And Development Of Complex Products and Systems" Lauren Aquino Shluzas, Martin Steinert, and Riitta Katila explore the pathways of user interaction in designing and developing complex products and systems. Two-phase research explores user involvement in early stage design and new product development.

Thomas Beyhl, Gregor Berg, and Holger Giese deal with the topic of bringing designers and engineers together in mutual understanding. "Connecting Designing and Engineering Activities" is about the use of artifacts of the digital or analog world.

2.7 *Part IV: Make Ideas Tangible*

The beginning article in this part delivers innovative companies who are adopting design thinking methods and techniques in order to try to create products with maximum end user value. The text focuses on small and big software companies who combine agile development processes with design thinking and examines the advantages of this approach. "A Research Plan for the Integration of Design Thinking with Large Scale Software Development Projects" is by Thomas Kowark, Franziska Häger, Ralf Gehrer, and Jens Krüger.

The authors of the article "Sharing Knowledge through Tangible Models – Designing Kickoff Workshops for Agile Software Development Projects," Markus Guentert, Alexander Luebbe and Mathias Weske, look at precise descriptions of the systems to be built. They describe that all data of a systems is compiled from observations and follow-up discussions with the user.

Bastian Steinert and Robert Hirschfeld wrote "How to Compare Performance in Program Design Activities: Towards an Empirical Evaluation of CoExist". It explores the design of an empirical experiment that compares programmers' performance in program design tasks. The target is to focus on the benefit of using CoExist in programming and accordingly make changes from within.

The last article shows how design thinking works in large companies. Holger Rhinow and Christoph Meinel take a closer look at management's perspective by an interview process in "Design Thinking Expectations from a Management Perspective".

3 Summary

Creativity and innovations need special relations and environments to grow, in other words a special innovation eco-system. It's not a process where pressure will foster results. It's a process where multiple approaches, mixed teams, and a lasting method may lead to success. Yes, innovation eco-systems need multidisciplinary teams, open-mindness, and different approaches. This new paradigm is hard to adopt by established institutions but it forms new approaches in old fashioned every-day challenges. The articles presented in this volume tell about different aspects how to follow the fundamental rules of design thinking in our design thinking research approach. These articles and results have their space, their fertile eco-system, and at last their time.

Design thinking is key to a modern world, a world where we can better get to know its challenges, where we can even better know how to tackle these challenges and to change the world into a better place. So design thinking is not only a nice idea, or a unique set of methods for everyday challenges, but it's a way of thinking about how to change the world. This is because of the way of thinking. The challenge may be unclear, the way to the solution also unclear, but we know that there has to be a solution because of the challenge before us. This challenge is connected with our environment of space and time, connected with our open mind, flexibility, the interest in solving wicked problems, combined with methods and techniques described in this book. We invite you to share our knowledge, walk in our shoes, and to feel attracted by this method called design thinking.

We'd like to thank all of the people who worked hard on this book. First of all authors and editors, but there have been even more people who worked on this volume: Martin Steinert who coordinated the research work of the participants from Stanford, Lutz Gericke who collected the articles of this volume and finally Sebastian Leder for all of his work in preparing this book and supporting its editors tremendously. But of course there have been many more people carrying the load who are not mentioned but not forgotten.

And at last we want to invite you, our reader, to our "Electronic Colloquium on Design Thinking Research" (ECDTR) on the internet at ecdtr.hpi-web.de where you can find many more interesting repeats, these and other material about Design Thinking Research. The ECDTR "is a forum for the rapid and widespread interchange of ideas, methods, and results in Design Thinking Research. The purpose of this forum is to use electronic media for scientific communication and discussions in the Design Thinking Research community." And it would be very interesting for you, us and our common discussions if you could share your ideas, methods and experiences on Design Think in this forum with us.

We hope that you will like our articles and our intention put in this book and that it's possible to start a dialogue with you on the use of Design Thinking in forming a smarter world.

Student Teams in Search of Design Thinking

Shelley Goldman, Zandile Kabayadondo, Adam Royalty, Maureen P. Carroll, and Bernard Roth

Abstract The research explored student teams as they worked independently of instructors and coaches to understand how students learn the design thinking process. Two approaches to the research were explored: taking cues from team members' reflections on their working sessions; and, analyzing communication bids made by students using interaction analysis techniques. Teams from two design thinking classes at the Hasso Plattner Institute of Design (d.school) at Stanford were studied. Results indicate that groups struggled for sustained and focused talk and activity relating to their assigned tasks, yet ultimately, established ways to communicate and accomplish assigned tasks. The findings implicate course design, suggesting more attention to team process and communication.

1 Introduction and Background

The movement to teach design thinking in universities is in full swing. A quick search reveals that, in over 60 universities and colleges, design thinking is taught as workshops, supplemental training, courses, or degree programs. These programs are in addition to the teaching of design as part of engineering, architecture, and art programs. The growing enthusiasm for teaching and learning design thinking raises questions about how this complex set of ideas, processes and concepts can best be taught. The field needs a better understanding of what happens in existing courses in order to improve both teaching and learning.

Design thinking is complex. It is about concepts, processes and the development of dispositions that guide thought and actions in innovative problem solving. This complexity poses several dilemmas for course-based design education. What can be

S. Goldman (✉) • Z. Kabayadondo • A. Royalty • M.P. Carroll • B. Roth (✉)
Stanford University, Stanford, USA
e-mail: sgoldman@stanford.edu; zandile@stanford.edu; adam@dschool.stanford.edu; carrollm@stanford.edu; broth@stanford.edu

H. Plattner et al. (eds.), *Design Thinking Research*, Understanding Innovation,
DOI 10.1007/978-3-319-01303-9_2, © Springer International Publishing Switzerland 2014

taught and what do we expect students to learn when we teach them design thinking? If the aim is to teach design thinking concepts, processes, practices, and dispositions, how do we set goals for what we expect students to learn, and how do we understand or assess their learning experiences? Should instruction focus on individual or team experiences?

Rittel and Webber (1973) used the term "wicked" to describe design problems. Cross (2006) extends that view to design thinking instruction, considering it as problematic as design is itself:

> It is also now widely recognized that design problems are ill-defined, ill-structured, or 'wicked' (Rittel and Webber 1973). They are not the same as the 'puzzles' that scientists, mathematicians and other scholars set themselves. They are not problems for which all the necessary information is, or ever can be, available to the problem-solver. They are therefore not susceptible to exhaustive analysis, and there can never be a guarantee that 'correct' solutions can be found for them. In this context a solution-focused strategy is clearly preferable to a problem-focused one: it will always be possible to go on analyzing 'the problem', but the designer's task is to produce 'the solution'.

In design thinking classes, instructors are trying to teach the components and process of design thinking, and also teach about the end game—transformations in the way the newly minted design thinkers think, act and intuit as they design novel and innovative solutions to problems. These ultimate educational aims set a high bar with observable and documentable outcomes, *and* with more subtle changes that are difficult to anticipate, define and document. von Thienen et al. (2012) liken the process of a group learning design thinking to engaging in group therapy and suggest that some of the processes, dynamics, and outcomes are comparable. Teaching and learning design thinking is a complex enterprise.

In teaching design thinking, courses aim to provide students with group learning experiences such as "radical collaborations," interdisciplinary experiences, and learning from deep exchanges with peers. Generally we teach students how to do design thinking and how to do it in teams. Instructors hope students have a productive team learning experience and, as such, rely on student teams to be proficient enough to carry the students through the process and projects that are assigned. The team process and practice is one of the sticky problems of design thinking education because courses are situated in educational systems that have emphasized and rewarded individual learning and achievement. It is not surprising then, that the team experience is in conflict with the individual achievement imperative, further complicating how we teach design thinking. Creating an effective and successful team learning experience is a sticky *wicked* problem.

As design thinking instructors and researchers, we aim to better understand the team learning experience and to find ways to better support it. When we began this research, we thought we would learn how to assess team learning in a design thinking course; when we finished, we learned that the teaching process might better address the student team experience. We were especially focused on the experience of groups during their out-of-classroom-time experiences, so we were able to capture how teams worked through the design process independent of their course instructors or coaches. The research taught us about how teams handle the design thinking process as they are learning and enacting it. If class time was when students were introduced to design thinking processes and mindsets, team meetings were times to fulfill assignment tasks in practice and production runs. They were also the times when important components of design processes and solutions – point of view statements, brainstorming, and design solutions – were experienced and negotiated.

We set out to assess team learning by starting with the basics and examining what the student teams were doing and how they were interacting. Previous studies were of interest. Kress and Schar (2012) examined cognitive differences among group members, and Brereton and colleagues (1996) determined that team interaction affected design outcomes.

Conflicts among group members seem endemic in teamwork and surfaced in this study. From prior research we know that a sense of belonging and togetherness, and sharing joint goals are important to design group's abilities to apply itself to its class projects successfully (Mercier et al. 2009). Teams worked best at enacting and representing what they were learning when individual members could not accomplish the tasks alone. Divide and conquer did not work best on design teams. Katu contends that, "Harmonizing is about emphasizing differences together" (Katu 2012, p. 18), suggesting that the best functioning teams succeed at keeping the team together despite members' differences. In his account, the effective and successful team members share passion, common goals, and commitment to excellence.

We examined student teams using methods enabling us to capture and take an ethnocentric view. Specifically, we followed two teams, from two courses, as they met outside of class to work on their class-assigned projects. Each group had 2–3 weeks to work on their own.

Together, the two groups show a process of students becoming introduced to design thinking and working to become acceptably proficient at it. Our results indicate that teams did not necessarily stick to the tasks that corresponded to their immediate assignments or their current stage in the design process. For example, a team with the object of prototyping, often drifted back and forth to earlier and later stages of the process.

We decided to focus on student teams in their independent work outside of the classroom. We acknowledge the facilitator role instructors play in design teams. As instructors attempt to create environments where teams can accomplish their independent work and achieve success, they are the supporting cast to the design thinking team ensemble. This study focused instead on students and their emergent roles in this ensemble. Both teams accomplished moments of unease and eventual alignment, illuminating the dynamic nature of teams in collaboration. This research on student teams' collaborative experiences provides a more nuanced view of how we might design more effective courses, and seeks to answer several questions: What is the nature of team collaboration for new design thinkers? How does what we learn about student teams implicate how design thinking might be taught?

2 Research Methods and Analysis

As researchers of design thinking and team learning, we were guided by social views of learning and a theoretical rationale that is based in socio-cognitive views of learning. Vygotsky (1976 [1934]) described how opportunities to interact with others in a social environment are essential to learning. The human-centered focus of design thinking and the deep and radical collaborations that define the process provide a deeply social process for learning. Design thinking is an approach to learning that encompasses active problem solving by engaging with (Dewey 1916), and changing, the world. These perspectives lend themselves to analyses of team interactions.

Qualitative approaches and methods guided our data collection and analysis. Our exploratory studies took a student-centered, emic approach and were conducted with students who were in Stanford design thinking courses. We studied teams of students who were enrolled in two of Stanford's Hasso Plattner Institute of Design (d.school) classes: *Design Thinking Bootcamp* and *Innovations in Education*.

Our goal was to gain understanding of student teams as they engaged in design thinking work. Our focus was on how teams who were learning design thinking moved beyond the in-class collaborations to new, and possibly, shared conceptual spaces. Our general frame of interest included the nature of group process, catalyst events, cohesion and affirmation, and group dissension. We wanted to study teams as they engaged in the practices and processes of design thinking to understand how they put their classroom learning into practice. Our goal was to do that as naturally as possible, to minimize our researcher intrusion into the team, and to work from the perspectives of the team members to guide our analyses.

The data collection process was comprised of three main data sources. The first data source was videotapes of team meetings made by the group members setting up a stationery camera and letting it run throughout their sessions. The second data sources were "confessionals" by team members called in to a telephone number after their team meetings, and emails among the group that were shared with the research team. The third source were post-project interviews completed with students as an option if they did not want to make "confessional" calls by telephone.

We had several reasons for creating this research design. First, we were interested in seeing things from different participants' perspectives to understand the significance of what the team members saw as key events. We felt this would help us gain more grounded perspectives on a collaborative effort. Second, we felt that it would provide a sense of privacy for the students. Instead of the research team asking questions, students had an opportunity to reflect and respond at their own pace and in their own words. The third reason was that we thought we would be able to quickly follow up with relevant questions based on the students' responses. And fourth, we used the student reflections to point us to events or instances they thought were significant, providing us with directives for where a truly student-oriented analysis might be focused.

Fig. 1 The telephone voice prompts

Voice Message Prompts

How would you describe your team after this meeting?

Tell us about some interesting, troubling, surprising, or exciting aspects of your meeting.

What did you learn about yourself, your team, and your team progress?

The "confessional" data was extremely relevant, and we privileged it in selecting data for deeper analysis. The voice messages turned out to be a productive data source as they revealed meanings students were attaching to the experiences they had with their teams. After each session, individual group members called a Google Voice number and responded to a set of prompts (Fig. 1). The three prompts were designed to keep reflections broad while helping students organize their thoughts. The phone messages were recorded as digital audio files and auto-transcribed with Google Voice software. We retrieved the files and manually corrected any mistakes in the transcripts. Excerpts from transcripts of confessionals (see Figs. 2 and 3) show the kind of reflections that were shared, including reporting on the context for the meetings, conflict that arose, personal reactions to events, and frustration and excitement.

Megan and Ellie's phone responses (Figs. 2 and 3) were made after the same team meeting. Both were members of Team One and point to a moment of conflict between teammates at one particular meeting. Together, these responses provided a focal point for our video analysis. In general, Team One appeared to have unresolved frustrations that members aired in their voice messages. Their process of working through team issues resembled "confessionals" in which the inner workings of the team were revealed to outsiders. (Recall how confessionals are overlayed on video footage in documentaries and reality television shows). These monologues, although external to the team collaborative process, revealed the more internal functions and dysfunction of the team that teachers and coaches might not readily have access to. We wanted to investigate how these moments of conflict had arisen and how, if at all, teammates had worked together to resolve it. We believed this kind of analysis would also indicate elements of individual behavior during team collaboration that foster synergy or disruption.

Team One had more moments of conflict in their team meetings, yet they bigger risks and pushed their prototypes beyond the boundaries perceived by individual members. Unlike the confessional feel of Team One's communication with the research team, Team Two's reflections were more spirited. In Fig. 4, we excerpt some comments from Nora that illustrate how team spirit was built and maintained primarily through email exchange.

While the team members' comments portrayed a vibrantly positive team spirit, they were also often irresolute, with members expressing how well things went but being a little less certain they'd accurately perceived and characterized their team's

Fig. 2 Excerpts from transcript of an exemplary student phone response

> Transcript 11.15.11 9:51PM
>
> Hi. This is Ellie from Boot camp. Um, let's see. So we just had our group interview, um, first session. And it really went, really well! Lots of good ideas, it felt like we had fun. I'm definitely trying to be cognizant of, you know, not interrupting each other and, like, hearing our full ideas out, umm.... [00: 29] Sometimes... I guess I think one of my team members can be, can, like, shoot ideas down before hearing them out. So it's a little frustrating to me and I definitely am aware of the effect that has on me—you know, in terms of actively contributing and continuing to when you feel like you idea wasn't heard

performance. They had nothing to confess. In other words, members talked about enjoying the process and their team while having no sense of whether or not their teammates might agree. Team Two shared very extensive email exchanges with the research team. While both teams struggled to schedule times to meet, Team Two split up tasks as indicated in the email exchange in Fig. 5.

When asked about her views on perfectionism during her exit interview, Nora described the internal shift she had to make to adjust to her teammates' outlook on the design work at hand:

> So I definitely sort of let go between DP1 [Design Project 1] and DP2 [Design Project 2]. Like in DP1 I was highly perfectionist and kind of freaking out and frustrated and
>
> Just being like (to DP1 team members), 'You guys, we have to have a perfect final product.' And our final presentation wasn't perfect. There was just a lot of frustration or just lack – just total lack of communication. It was just so frustrating because I had invested all of this energy and then the output wasn't what I would have wanted it to be. So how I dealt with that going into DP2, I think, was putting less of myself into it so I had a lot less to lose. But not in a way that I completely divorced myself from this process, but sort of in a way that like, 'I'm going to be more balanced in my approach to this, and rely on my team members and if they don't do things the way that I would want to do it then that's okay.' You know, so striving for my own version of perfectionism less.

During this exchange, Nora, explains her team's interpretation of "embracing" design thinking as learning to let go of perfectionism, but misrepresents the potential gains inherent in a more critical approach to her teammates contributions. This meant that, a large part of their work was done individually with teammates later reviewing each other's progress and offering revisions. This approach to collaboration presented fewer opportunities for teammates to challenge each other, and influenced their team meetings (where all members were present) accordingly.

One important difference between the teams was the stage of the design process captured in the videos we analyzed. Recall that with Team One, we zeroed in on a moment of crisis indicated by a team member. In Team Two, we looked for a

Fig. 3 Sample transcript from a student phone response

Transcript 11.16.11 5:41 PM

Hi, this Megan from Team One. So we had a meeting earlier today with all of us, and we were going to do a prototyping session which I thought ended up pretty successful. We kinda didn't have all that much time so we really got to it and stayed focused.

[00: 27] One thing that I, ugh, thought was interesting was that, at one point, two of my other team members kind of had a disagreement about the approach of one of the prototypes and got into a little exchange – not heated by any means, but kind of expressing their different points of view. And I actually really appreciated it because, um because I feel like I'm usually one of the more expressive and vocal members of any of the teams that I am on. And, so it was nice to kinda not be in that and to be an observer and to not feel, like, particularly strongly either way about the issue, which was different for me. And I think, I think a lot of times when I'm in a team setting, I try to pay attention to, like, how I can tone down—like, if I have a strong point of view—how I can tone it down and I just thought that it was interesting for me to kind of just be an observer in that situation.

[01.50] One thing that I feel like...I feel like our team's doing really well and we're together, we have productive meetings. Something that I think has been like a little bit challenging has been coordinating schedules and also, like, not everybody being every meeting… the ideal would be obviously, if all of us were there at the same time. And I just think that like a significant amount of time is rehashing and getting everyone on the same page. Another observation. Umm...

similar moment of crisis and its subsequent resolution. Understandably, the team meetings we chose for our analysis had disparate design agenda and team imperatives. Team One's meeting was closer to the beginning of the design process, with members struggling over the transition from empathy work to defining a Point of View statement (POV). The conversation and activity in this meeting occasionally drifted to other stages in the process (for example, George brings up a POV that has good implications for a future prototype), and involved very little explicit

Fig. 4 Excerpt from student exit interview

Nora Smith(NS): Yeah, so if you couldn't contribute to one part that was fine, but then you made up for it later so... And I think everyone walked away from the project...my sense is that everyone walked away from the project feeling pretty good about what they had contributed and what everyone else contributed. And it was like a pretty positive team dynamic in general. Like we would send around emails...I would send something out saying (smiling and miming typing), "I'm going to do this." And Steve would respond (miming typing), "Way to go, Nora!" And it was very, like, "Go Team!" And that really came from everyone, where we all sort of had a good team spirit—for the most part—moving through the project. That is good, that positive reinforcement: making other people feel really good about what they are doing, that worked out pretty nicely.
ZK (researcher): Did that come up organically because of the personalities or was there some sort of...norms that were set up in the beginning
NS: It was kind of weird. I think Steve actually was the one who initiated all the, "Way to go!" (laughing fondly) and the "Go us!"
ZK: How did it make you feel the first time he did it?
NS: It was great! It was also really unexpected because I also worked with him in DP1 [Design Project 1]. DP1 was just a very different project. We... it was me and Steve and Jane [another member of the class], and we kind of had our own... we had a lot of frustration with that process and project. And I think me, Steve and Angie...or Steve, Angie and *I* worked better together than Steve, Jane and I did. So having worked with Steve already and not having had that positive, [said enthusiastically] "Yeah! Way to go!" like that sort of attitude. So the first time he did it I was like [laughing], "Where is this coming from?"
ZK [laughing]: ..."Who is this person?"
NS: Yeah..."Who *are* you?" And you know, I thought maybe he was just having a particularly good day and wanted to send out a good email. But then it sort of...Angie and I started to pick up on that a little bit and it sort of became this *thing*, that in our email exchanges...we'd sign it like, "Go Team!" or whatever. [Laughter] Or just be like, "Nice work guys, I think we really came together well on this one." Just nice things like that. And when you have a team dynamic like that...and I think it's a little chicken-and-egg-y, like maybe you have a good team dynamic and inspires comments like that or maybe you have comments like that and that inspires a good team dynamic. In this sense, we did start with a good team dynamic and good individual contributions that inspired us to send around positively reinforcing comments. Not necessarily that the positively reinforcing comments inspired the good individual contributions.

Fig. 5 Sample email exchange between team members

----- Forwarded Message -----
From: Angie
To: Nora, Steve Cc: REDlab
Sent: Thursday, May 24, 2012 1:40:52 PM
Subject: Re: Improved prototype

Thanks guys see u all later

On May 24, 2012, at 12:30 PM, Nora wrote:

Hey guys,
Some updates:
I read through the booklet and made a few changes, but since it's just a prototype, perfection isn't a requirement. I think it's a good start.
I think it'd be great if we could photo-document our meeting this afternoon and maybe have the preceptors (and any other volunteers we can find around tressider?) roleplay…

On May 24, 2012, at 11:37 AM, Steve wrote:
Angie and Nora,
Here is an updated document. Primarily I spread the formatting out and made it bigger - I think we could make it more fun, but that would be another rev.
If you get a chance, go for it, otherwise I think it is good enough from my side. I want to hear more and ask questions about their expectations and hopes and reflections about being an incoming freshman that I want them to react.

On May 22, 2012, at 3:04 PM, Angie wrote:
Hi guys,

this is what i have so far for the booklet. Please feel free to make any changes and add anything that you want. Send me your version by the end of tonight so I can have a full prototype tomorrow before meeting the preceptors!

discussion of team collaborative processes and no talk at all of logistics. In contrast, the primary objective of Team Two's meeting was to more clearly articulate their prototype and a considerable amount of team talk and activity was dedicated to team collaborative process and logistics, with team members dividing up tasks to be completed and even performing these in the meeting. Angie, who was in charge of editing the team's presentation, the final deliverable for the class, often brought the conversation back to how they could capture the design process and document their insights.

3 Analysis

With the videos, confessionals and interviews completed, analyses were conducted to make use of the student provided information and directives. Analyses started with open coding of the confessionals and interviews. We triangulated across data sources, finding instances on the video that were reported in voice messages, emails, or interviews. The research team watched video segments and chose one from each team for analysis. The selection process involved trying "see" the issues students described for us and identifying a beginning and an end to the respective events to which they referred. Once segments of the group in a topical event were identified, we created a content log that described what happened throughout.

The preliminary analyses and the emphasis on team interactions led us to look at how the groups attended to and accomplished their interactions. Since we were already rooted in the idea that talk and action were some of the building blocks of group work and learning, it made sense for us to look at the task the group was working on, the movements they made in talk and related actions, and how and when their interactions were relevant to the design thinking process they were learning. We concentrated our analyses on when "bids" were made in the group and how they were received and acted on. Bids are a struggle for control, attention, or for the right to speak within a group (Schegloff 1998, 2007). We considered bids to be requests/imperatives for action, type of work, adherence to process, and attention to relevant or irrelevant topics. They were requests for action from the team members, and we expanded the definition to include both verbal and non-verbal requests. A bid was as simple as requesting a turn to talk or building on another's idea, or as complex as entering a new topic into the conversation or suggesting the design solution to the challenge. Because bids were invitations or requests for interaction, we looked at bids offered by the students and subsequent responses. When we began watching the first group, we realized that the students were having a difficult time staying with their task to develop a POV and appeared to be all over the process map, pulling anything they knew with respect to design thinking into their activity. We decided to analyze the students in terms of their talk, focus, and action, the topics they were taking up, and the design thinking skills or processes they were invoking in the moment.

4 Coding Categories

The research team developed a series of codes by which to analyze the video. We looked for verbal bids, non-verbal bids, and their responses in the interactional palette. The non-verbal included movements such as changing place in the workspace, grimacing, pointing, or writing on the board with a marker. By repeatedly watching the videos, we conducted an open coding process, allowing codes with corresponding numbers to be generated and defined. The codes were not

[Clip] Time	Ellie	Ellie Bid/Response (Verbal)	Ellie (Spatial)	Ellie (Objects of focus/Topics)	Megan	Megan Bid/Response (Verbal)	Megan (Spatial)	Megan (Objects of focu/Topics)	Megan bid short
0:00:19	[Crosses from bottom left to the table and uncaps a marker from the table and prepares to write] I just like that idea of like...	2	60		[Comes back onscreen. Wlks around table: from offscreen, along the left, to now directly facing camera. places white folder on the table and opens it]		30		1
			50						
					Yeah, that's actully really interesting	4	50		1
0:00:25	Like "Tv-Turn Off-Week"	4	10	103	[Pulls out folded chart which looks likes a spreadsheet. Chart rustles as George elaborates on the prototype]		50		3

Fig. 6 Segment with bid and response codes by number by team member

exhaustive of all possible codes, but did capture the range in these particular data. The codes established as *Verbal Bid Responses* are examples of this: "Acceptance" was defined as agreeing with the previous bid; "Building up" was defined as adding an additional action or idea to the previous bid; "Ignoring" was defined as not responding to the previous bid or responding with an unrelated action or idea. "Rejecting" was defined as explicitly disagreeing with the previous bid. Examples of *Spatial Bids* included, "Writes," "Proxemics moves," "Gestures," and "Attending to one's own person" (fixing hair, arranging clothing). *Objects of Focus* codes included "Design thinking steps," "Team collaboration," "Logistics," and "Social work." The preliminary code list was expanded and refined during the coding process.

From coding we realized that spatial and verbal bids and responses were being used to affirm or dissent in similar but distributed ways. Some team members would show agreement by leaning in and looking directly at their teammates when these teammates had the floor to speak. At other times, they would verbalize their agreement with "Umm, hmn" and "Sure! I think that's great!" In both cases the bids proposed by their teammates were *accepted* (Fig. 6).

In Team One, Ellie used her spatial responses for expressing dissent more often than other teammates: walking offscreen to a different whiteboard; opening a can of soda in response to a direct question; rifling through a stack of post-its while her teammates were looking over a chart together; these were all examples of how she *ignored* her teammates bids. These forms of ignoring and accepting were more nuanced and passive forms of dissenting and affirming, respectively. In Team Two, Angie was often offscreen, or documenting somewhat unrelated material on the whiteboard, or preparing for the next stages of the design process on her laptop.

Angie's teammates tended to overlook these dissenting spatial bids so that the bids interrupted the flow of ideas less often than Ellie's did. In Ellie's team, the team members on the receiving end of passive bids were highly attuned to them, stopping what they were doing in response to disruption of spatial synergy, and in these instances dissent appeared harder to resolve.

There were "idealized" and active forms of responding to bids, such as when a teammate responded to another's bid by elaborating on it. There were also passive forms that could be affirming, disregarding, or dissenting. When Steve copied down Nora's comments on the board, this was a move to *build-on* her points. When Megan responded to Ellie's bid with a counterargument this was a move to *reject* the bid made. Often, build-ons and rejections were complex and compounded, with spatial and verbal sub-bids embedded in them. For example, George, in response to a bid Megan makes, nods then pauses before adding,

"I think 'why' is more interesting...[Walks over to the board. Takes his time skimming it.]

Why is this the center of his life? [Approaches the circled "center of his life" and taps it with this finger]"

In this instance, the nod is a spatial acceptance. The comment, "I think 'why' is more interesting..." is a verbal challenge that modifies Megan's comment and shifts the team's focus in a dramatic way. This bid is a verbal build-on. When George walks over to the board, he offers a new spatial bid by demanding the attention of his teammates, who in turn accept his bid by following him with their eyes. He uses this to build-on his own question, "Why is this the center of his life?" and spatially builds-on what Ellie had written on the board "center of his life." With this compositeness in mind, we recoded the transcripts for both teams, this time condensing bids and responses to one of four categories: (1) accept and (2) build-on, the two affirming activities; and (3) ignore and (4) reject, the two dissenting activities. Figures 8 and 9 show the distribution of the four categories in Team One and Team Two, respectively.

Figure 7 indicates a fairly evenly distributed use of talk-space by the three members of Team One. George built-on bids more than he rejected, ignored or accepted them. Megan, rejected more than she built-on, accepted or ignored. Ellie rejected and built-on less than she accepted and ignored.

Figure 8 illustrates that while Nora contributed the most to the team talk space, she built-on far less frequently than she accepted, ignored and rejected her teammates bids. Steve on the other hand, took up the least talk-space and most of it was accepting and building-on (his distribution is widest for these two forms of bid response). Angie ignored far more than she accepted, but also built-on more than she rejected.

In the following series of graphs, we show how these patterns of individual behavior related to uptake of ideas. In our coding scheme, we cross-referenced the bids and responses of each member. For example, for all of Georges bids, we counted how many times either Megan or Ellie responded by elaborating, whether

Fig. 7 Distribution of bids in Team One meeting

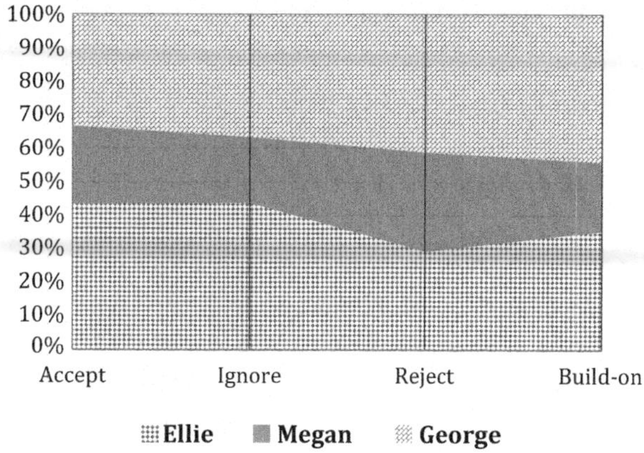

Fig. 8 Distribution of bids in Team Two meeting

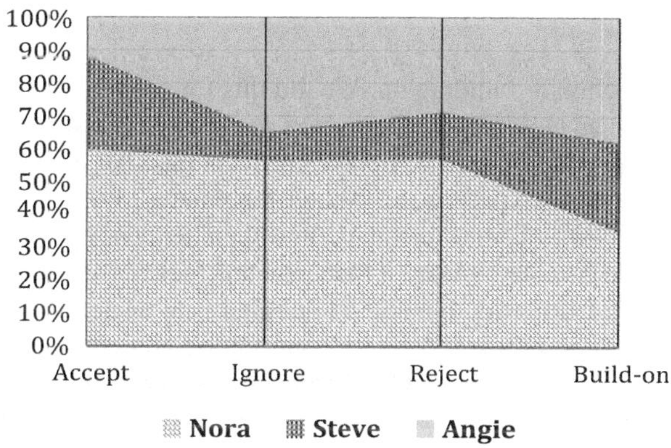

Fig. 9 Dissenting bids and uptake in Team One

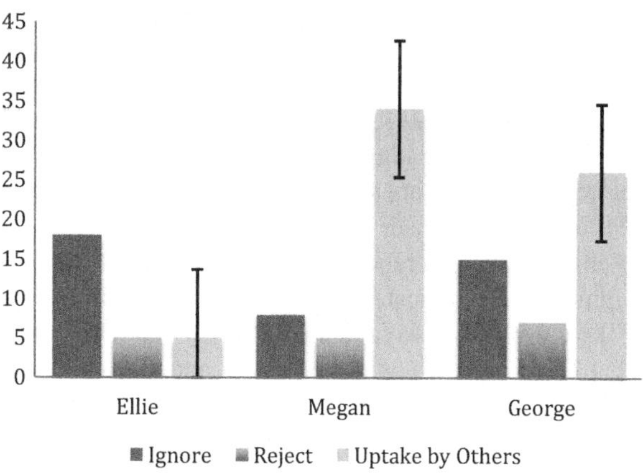

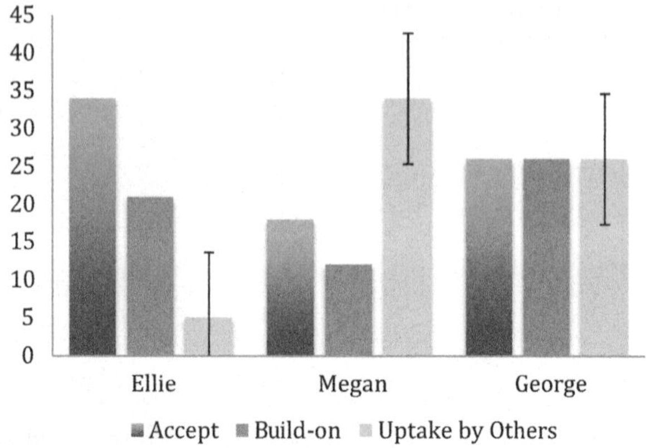

Fig. 10 Affirming bids and uptake in Team One

rejection or building on. We did this for each member of both teams, and came up with an *uptake* count that serves as a proxy for how *visible* that member was to their teammates. The more uptake one team member had, the more visible they and their use of talk-space were to their teammates. Was there a relationship between the visibility of team members and their tendency to affirm or dissent? Recall how Ellie's voice message influenced our foci for analysis. In Ellie's words, "So it's a little frustrating to me and I definitely am aware of the effect that has on me – you know, in terms of actively contributing and continuing to when you feel like you idea wasn't heard." This particular relationship between uptake (or visibility) and bid type would help us explore Ellie's claim about not being heard. Figures 9 and 10 show the relationship between uptake and the dissenting bids, and between uptake and the affirming bids for Team One. Each graph shows three bars for each team member: two bars for bid type, and the third for uptake count. In Fig. 9, Ellie, for example, ignored her teammates bids 18 times, and rejected them 5 times. Her teammates responded to these 23 combined dissenting bids only 5 times.

For Megan, who ignored teammates's bids 8 times and rejected them 5 times (a total of 13 dissenting bids), team members responded (by accepting, building on or rejecting) 34 times. George's total of 22 dissenting bids were responded to 26 times. Megan had, by far, the most uptake.

In fact, the difference between Megan and Ellie's uptake counts was statistically significant, as indicated by the Y error bars on Figs. 9 and 10. This indicates that Ellie's visibility was radically different and less than Megan's and George's during that team meeting, while Megan's and George's visibility was not significantly different from each other. Ellie was right! She wasn't being *heard* as much as Megan or George were. We mapped the relationship between uptake and type of bids for Team Two but the results were not significant, indicating that each member of the team was being heard (or had their bids taken up) by roughly the same amount.

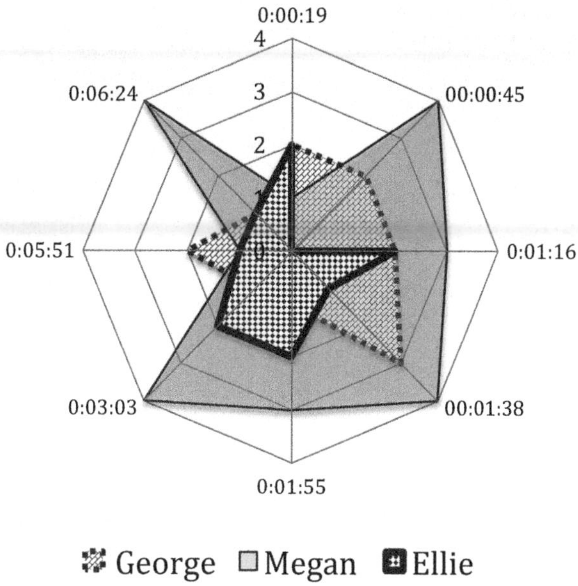

Fig. 11 Tension and alignment in Team One

We suspected this difference in visibility had something to do with how the teams dealt with dissent and affirmation or, more specifically, how they transitioned from one to the other. We identified one moment in each team where conflict arose, where all these members were spatially present, and where some resolution seemed to be achieved. In Team One, this moment arose when there was some dispute over who the user was and what his explicit and implicit needs were. In Team Two, the moment arose when one team member, Nora, suggested discarding a key component of the team's final product, the ePortfolio. In Team One, this moment occurred at the beginning of the meeting, from 0:00:19 to 0:06:24; whereas in Team Two it occurred at the end, from 0:37:17 to 0:44:10, with the meeting and video recording ending 30 s later. To examine this in more detail, we divided these portions of the meeting into eight smaller time segments, each corresponding to a bid made by one team member and the two corresponding responses to it from his or her teammates.

The two radar graphs, Figs. 11 and 12, indicate the flow of bid-making and bid-responses that took place over those eight time segments. These snapshots in time show the transition from bids that accept and build-on to bids that ignore and reject team members' contributions. Our data shows one team (Team One) is overly erratic and one team (Team Two) that is more linear. The radar graphs illustrate what moments of tension and moments of alignment might look like for both teams through the lens of offering and accepting bids. When bids are easily accepted and built upon, the team moves smoothly and cohesively. The bid-making patterns of

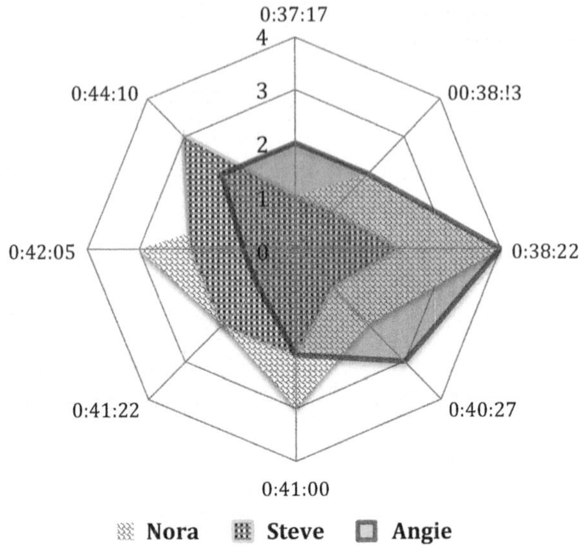

Fig. 12 Tension and alignment in Team Two

individual members resemble each other or are closely *aligned* when members take equal turns to lead, follow, challenge and affirm team progress. In an aligned team, even as members disagree with each other, there is synergy. The flow of the team is dynamic and emergent and is not easily attributed to one member but rather to how they interact and act in concert. In contrast, moments of tension in an erratic team are represented by scattered bidding patterns. Here, members' contributions are misaligned and the shifts from affirming bids (bids that accept and build on) to dissenting bids (bids that ignore or reject) appear abrupt and disjointed. Members appear to be working individualistically.

Figure 11 depicts Team One's struggle to find synergy. The team member's use of affirmation and dissent, passive and active, verbal and spatial are radically different from each other, indicating a struggle to come together. At time 0:00:19, two members are building on (level 2 on radar) while one accepts (level 1 on radar). This is "coming together" but in the next moment, time 0:00:45, one member rejects (level 4 on radar) while the second ignores (level 3 on radar), and the third checks out by wondering offscreen and out of the team space (indicated by 0 on radar).

Below we include the transcript for Team One that corresponds to these eight time segments. Each time segment comprises one bid and two subsequent responses that (1) accept, (2) build-on, (3) ignore or (4) reject the bid in question. Time segments 0:00:45 and 0:03:03 show mixed responses to bids, misalignment and disarray; while time segments 0:01:16 and 0:05:51 show team alignment: somewhat strong and very strong responses to bids.

At 0:00:45 Team is misaligned mixed responses to Megan's bid

Megan	[Consulting the spreadsheet] He's not on this list.
George	[Leans in to see the list. Pauses. Leans closer and points to a line on the spreadsheet repeating the name:] "Jeff" [row] "22."
Ellie	Oh! Here we go! [Talking to herslf. Finds the spot she's looking for on a second whiteboard and strides across the room to place the post-it she's holding on it disappearing offscreen.]

At 0:03:03 Team is misaligned mixed responses to Ellie's bid

Ellie	And also he's not getting exercise!
Megan	[Biting on the cap of the marker] Hmm. But he seemed pretty fit to be honest. He had kind of a beer gut [Chuckles, taps marker with her finger] He didn't seem... He seemed...Like, he wasn't like this... [flicks head to one side]
George	[Ignoring question consults the chart again].

At 0:01:16 Team is aligned somewhat strong responses to George's bid

George	Ok now can we sum [gestures "bringing together"]...[continues to speak but cut off by Ellie]
Ellie	[Picking up from George] I think we've hit a couple of different "needs"
Megan	[Picks up a marker from the table and spins around to face the onscreen whiteboard. She uncaps the marker and prepares to write]. Well let's maybe may try to figure out what they need...

At 0:05:51 Team is aligned very strong responses to bid

George	He has these interactions they are not... [pause as he thinks] ...they are either too short to be meaningful, or they are like...
Ellie	[Perks up and gestures to Megan. Scratches nose while looking at George. Leans forward with elbow on knee. Presses fingers to lips and looks in George's direction]
Megan	Umm hmm

Figure 12 illustrates a smooth transition from tension and dissent to alignment and affirmation over 5 min of Team Two's meeting. The team disagrees in the first few minutes and finally arrives at a moment of synergy. One member follows the other's lead, navigating the problem solving process in a fluid fashion. For example, at time 0:38:13, two out of the three members are building on (level 2 on radar) an idea, they quickly shift to rejecting bids (level 4 on radar), and then to ignoring (level 3 on radar) some and finally building on again at time 0:41:22. While this team has synergy, it does not move in a straight line, and these moments of tension allow the team to discover interesting ground. That this team accepts a lot of bids without building on them or disagreeing helps them interact smoothly, but they don't take any risks or take up many novel doesn't or creative ideas. Below the graph we include a segment of transcript for Team Two that corresponds to a several of these eight time segments. Like the time segments for Team One, each comprises one bid and two subsequent responses that (1) accept, (2) build-on, (3) ignore or (4) reject the bid in question. Time segments 0:38:22 and 0:40:27 show weak response to bids, misalignment and disarray; while time segments 0:41:22 and 0:42:05 show team alignment: very strong and somewhat strong responses to bids, respectively.

At 0:38:22 *Team is misaligned, weak responses to Steve's bid*

Steve	If the technology works and this does transform that space then we'll take that on. I'll take that on as a primary...
Angie	[Interrupting Steve but addressing everyone] Do you think all these are fun? I'm not sure how much fun that part of the

	program... I'm not sure if the game should be focused on helping the students or...
Nora	I think that any of these games could be framed in a fun way, or framed in a more informative way. I think they can be fun and incredibly important.

At 0:40:27 ***Team is misaligned, very weak responses to Steve's bid***

Steve	You were asking [addressing Nora], "What do you mean by: Games with No Rules so that People Can Create Them?" And one game that I was thinking about was–have you ever played [chuckles] Quarters?
Nora	[Ignoring question] Building off of this [leans in and point to "friendly competition" on the board], the "friendly competition", the "one-ups manship" and all of that, umm...I think that plays into games like, "Yes, and!" and plays into games like the Incense Story(?) where people say: [starts playing the game]:
Angie	[Walks towards the teacher – who has just poked her head around the team's whiteboard to check their progress – but the teacher has already walked away from the group].

At 0:41:22 ***Team is aligned, strong responses to Nora's bid***

Nora	So we have to sort of harness it and direct it [gesturing her meaning] in a positive direction through these games is a good thing.
Angie	Are we...[pauses as Steve talks].

Steve	That's a really good thing to capture that thought [Steps in to write on board: "harness competition"]

At 0:42:05 *Team is aligned, mixed responses to Steve's bid*

Steve	I mean. . .[struggles for words] I would personally wave at the Fall, kind of like say a vision but not create it. It's too big to actually poof out.
Angie	. . .right. . .
Nora	[Rubbing her lip as her eyes continue to wander]

5 Summary, Discussion and Implications

The goal of the *Assessing Team Learning Project* was to gain insights into how independent team collaborations were accomplished as students engaged in and learned design thinking. The ethnocentric methods we employed allowed for deep insights into the nature of the process. Through taking an approach to data analysis that privileged our study participants' views of their experiences of their groups, we were pointed to particular events for study in videotapes of team meetings. We started the analyses by looking at what the students reported as problematic or extremely interesting or satisfying. Central to our analysis was the idea of a "bid" for topic introduction, change, or redirection of focus. A bid could be verbal or non-verbal. We also identified responses, from ignoring bids, entering competing bids, building on bids, etc., and traced the trajectories that bids and responses took in the group interactions. Within the larger theme of team interaction, we gained knowledge about how bids and their responses impacted on how teams functioned to accomplish their goals. We explored the diverse interactions that students engaged in, how they negotiated their bids to make contributions, and the effects of their participation moves.

6 Findings

Several findings emerged from the analyses: Team One touched on many stages of the design thinking process [empathy, define, ideate and prototype] in working to find its way. This might have been amplified because the students had trouble connecting to each other or immediate aspects of the design task (development of the POV). This also might have been their first meeting independent of class. The team established uneven participation patterns and those patterns resulted in them noticing tensions. Ellie's report that she was not being heard was validated by the analysis.

One lens for understanding the data, especially the bids, was the examination of team direction and alignment. One might think that a good collaborative design team navigates a problem solving process smoothly, and that a team where members are not on the same page and/or are abruptly changing directions might fall prey to ineffective problem solving and run of the mill design solutions. Through the analysis we generated an *uptake* count that served as a proxy for how *visible* that member was to their teammates. When we first started, we thought the more uptake one team member had, the more visible they and their use of talk-space were to their teammates. We found this to be a false assumption. In fact, the team that moved in strict linear manner and had shared uptake on bids did not discover any novel ideas or take risks. The group that seemed less focused and experienced the most unevenness ultimately found ways to negotiate their interactions and get on track with each other. Team One had a rougher time than Team Two, but may have had more overall success in designing a creative and innovative solution. The conflicts were not necessarily unproductive in relation to end-results.

In Team Two, the students contributed fairly evenly or equally in the collaboration. Their "even" style, while conflict-free and less frustrating, misrepresents the potential gains inherent in a more critical or uneven approach. The team had more convergence and less conflict, and the results of the analysis of bids and uptake were not significant. Ultimately, a large part of their work was done individually with teammates later reviewing each other's progress and offering revisions. They had fewer opportunities for teammates to challenge each other, and they may have sacrificed innovation in their class project.

When bids were accepted, built upon, and subject to interpretations, the teams moved smoothly and cohesively. The bid-making patterns of individual members resembled each other or were closely *aligned* when members took equal turns to lead, follow, challenge and affirm team progress. In an aligned team, even as members disagreed with each other, there was synergy. The flow of the team was dynamic and emergent and was not easily attributed to one member, but to how they interacted and acted in concert. The data showed, in contrast, that moments of tension in a team were represented by scattered bidding patterns. Members' contributions were misaligned and the shifts from affirming bids (bids that accept and build on) to dissenting bids (bids that ignore or reject) appeared abrupt and disjointed. Members appeared to be working individualistically in these times of tension.

7 Implications

Much work has gone into designing course pedagogies at the d.school, where this study took place, and our team's observations of five or six courses revealed that courses and pedagogy were in line with the standards called for by design education researchers (Brereton 1996; Dym et al. 1999; Dym et al. 2005; Gerber 2009). These standards included: using problem-based learning and other appropriate pedagogies, making the investment in instructors and group coaches, considering it a crucial investment to educate diverse students with and into design thinking. Attention was paid to developing projects, in class activities, reflective practices (Adams et al. 2003), and assessments. Teams had instructors and groups have coaches. While students were in class, they were guided to engage the design process and mindsets. They were also guided to engage in "idealized" ways, receiving instruction and practice on different aspects of the design process. For example, when they covered how to write a Point of View statement in class, the very task Team One had trouble with, they received instruction and completed activities that included: what the POV is, why it is important, what qualities and standards a POV should meet, and ways to check their POV in order to ensure it is an adequate one. They learned about, and practiced writing POVs in class. Still, our video revealed Team One floundering as it worked independently to develop a POV statement for its user. Even when instructors work extremely hard on course design and cover so many bases, the teaching and learning of design thinking has its *wicked* aspects, such as the team collaboration.

The results raise questions and suggest implications for teaching design thinking and the need to better support independent teamwork.

First, is important to determine how to best help teams manage the design thinking process as they move through the different stages of a project. Finding ways to attend to team interactions in the design thinking process may pay off in terms of groups' overall experiences and success in generalizing solutions.

It is important to pay attention to teams' abilities to recognize ambiguity in the design process. In a prior studies to this one, our research group found that students did not become design thinkers in a developmental sequence. Instead, there were moments of significant insight that shifted one's understanding of the mindsets and processes that underlie design thinking (Goldman et al. 2012).

The development and handling of "teamness" is significant and worthy of extra attention. Teams are comprised of students with different backgrounds, disciplines, and prior design and team experiences. These differences bring both advantages and possibilities for radical collaborations (Booker et al. 2009; Barrick et al. 1998), and students need pointers about how to manage, massage, and capitalize on their differences in support in the instructional process. Students may benefit from the introduction of varied kinds of analytics in the design thinking process such as the creation of team rubrics and specific reflections on team process. Focusing on how teams collaborate in design thinking might benefit from a greater emphasis on evaluating the team process rather than just the end of course design solutions.

John Dewey (1922) wrote, "Conflict is the gadfly of thought. It stirs us to observation and memory. It instigates to invention. It shocks us out of sheeplike passivity, and sets us at noting and contriving." Design thinking relies on the resolution of conflict between a sticky problem and an elegant solution, what is known and unknown, what end-users say and what they really mean, and what does and doesn't work for users. There are times where novice design thinkers are asked to make inferences about people, their needs, the possibility for solutions that will work, and what will pass muster in terms of grading of their work. There is no wonder why student teams seem unanchored in the design thinking process when they work independent of their instructors. The teams we studied found their way, more or less, and presented solutions that met course criteria. Internally, their team processes were not elegant, and they stumbled through and *around* the design process. Some of what they were being taught proved useful in helping them become more attuned and responsive to each other as team members. The process was not conflict free when they worked outside of class, and both groups struggled to achieve a level of "teamness" that enabled them to accomplish their course and project goals.

Acknowledgments We would like to thank the students who participated in our research. They have contributed to our understanding of how instruction might better impact their learning. A grant from the Hasso Plattner Design Thinking Research Program made this work possible. Findings and opinions presented are those of the authors and do not represent the program.

References

Adams RS, Turns J, Atman CJ (2003) Educating effective engineering designers: the role of reflective practice. Des Stud 24(3):275–294

Bao P, Gerber E, Gergle D, Hoffman D (2010) Momentum: getting and staying on topic during a brainstorm. In: Proceedings of CHI 2010, ACM Press

Barrick MR, Stewart GL, Neubert MJ, Mount MK (1988) Relating member ability and personality to work-team processes and team effectiveness. J Appl Psychol 83(3):377–391

Booker A, Goldman S, Mercier E (2009) Interdisciplinarity in learning technology. In: di Giano C, Goldman S, Chorost M (eds) Educating learning technology designers: guiding and inspiring creators of innovative educational tools. Routledge, New York

Brereton MF, Cannon DM, Mabogunje A, Leifer L (1996) Collaboration in design teams: how social interaction shapes the product, analyzing design activity. In: Cross NG, Christiaans HHCM, Dorst K (eds) Analysing design activity. Wiley, Chichester, pp 319–341

Cross N (2006) Designerly ways of knowing. Springer, London

Dewey J (1916) Democracy and education: an introduction to the philosophy of education. Macmillan, New York

Dewey J (1922) Human nature and conflict. Holt, New York

Dym CL (1999) Learning engineering: design, languages, and experiences. J Eng Educ 88 (2):145–148

Dym CL, Agogino AM, Eris O, Frey DD, Leifer L (2005) Engineering design thinking, teaching and learning. J Eng Des 94:103–120

Gerber E (2009) Prototyping: facing uncertainty through small wins. In: Proceedings of ICED

Goldman S, Carroll MP, Kabayadondo Z, Britos Cavagnaro L, Royalty AW, Roth B, Kwek SW, Kim J (2012) Assessing d.learning: capturing the journey of becoming a design thinker, with. In: Meinel C, Leifer L, Plattner H (eds) Design thinking research: measuring performance in context. Springer, London, pp 13–33

Katu (2012) Let's spend our lives together. Bus Life 16–22

Kress GL, Schar M (2012) Teamology – the art and science of design team formation. In: Plattner H, Meinel C, Leifer L (eds) Design thinking research: measuring performance in context. Springer, London, pp 189–209

Mercier E, Goldman S, Booker A (2009) Focusing on process: evidence and ideas to promote learning though the collaborative design process. In: di Giano C, Goldman S, Chorost M (eds) Educating learning technology designers: guiding and inspiring creators of innovative educational tools. Routledge, London/New York, pp 36–61

Pimmel R (2001) Cooperative learning instructional activities in a capstone design course. J Eng Educ 90(3):413–421

Rittel H, Webber M (1973) Dilemmas in a general theory of planning. Policy Sci 4:155–169 [Reprinted in Cross (ed) Developments in design methodology. Wiley, pp 135–144]

Rowe PG (1987) Design thinking. MIT Press, Cambridge, MA

Schegloff EA (1998) Reply to Wetherell. Discourse Soc 9:413–416

Schegloff EA (2007) Sequence organization in interaction: a primer in conversational analysis, vol 1. Cambridge University Press, New York

von Thienen J, Noweski C, Meinel C, Lang S, Nicolai C, Bartz A (2012) What can design learn from behavior group therapy? In: Meinel C, Leifer L, Plattner H (eds) Design thinking research: measuring performance in context. Springer, London, pp 285–302

Vygotsky LS (1976 [1934]) Thought and language. MIT Press, Cambridge, MA

Team Cognition and Reframing Behavior: The Impact of Team Cognition on Problem Reframing, Team Dynamics and Design Performance

Greg Kress and Joel Sadler

Abstract As designers collect information about a problem, they form a mental frame of the problem space that is the scaffolding around which to build a solution. When presented with new information, successful designers can "reframe" the problem and the solution as part of a successful iterative cycle. These iterative cycles are central to the Stanford Design Thinking process. However, individuals and teams reframe to differing extents; is this variation rooted in intrinsic differences in cognitive style, and can it be associated with long-term innovative performance? We propose and evaluate a closed-form assessment tool called the Stanford Design Thinking Exercise (SDTE) to answer these questions. The results shed light on the particularly strong need for improved team dynamics measurements and the challenges of transcending context-specificity. Pathways for enhanced team dynamics measurements are explored.

1 Introduction

An increasing amount of work takes place in a team context, and there is a need to understand what factors can effectively diagnose, facilitate and predict team performance (Skogstad et al. 2009). Prior research has shown that certain team dynamics indicators are strong correlates with long-term innovative performance (Jung et al. 2012; Kress and Schar 2011; Tang 1991). There is also evidence that these dynamics may be the result of underlying intrinsic compositional characteristics (Wilde 2008). However, current team observation techniques are generally time-intensive (e.g. direct observation in the field) and often substantially asynchronous (e.g. offline video analysis) or otherwise obtrusive to team function (Tang and Leifer 1991). Another major challenge is that the definition of team

G. Kress (✉) • J. Sadler
Stanford University Center for Design Research, 424 Panama Mall, Stanford, CA 94305, USA
e-mail: glkress@stanford.edu; jsadler@stanford.edu

H. Plattner et al. (eds.), *Design Thinking Research*, Understanding Innovation,
DOI 10.1007/978-3-319-01303-9_3,

performance is highly context-specific; therefore, a model is desired that has demonstrated validity across different team contexts. In particular we are concerned with understanding those factors which are reliably predictive of team performance. We have identified and explored three lines of inquiry toward developing a solution to this problem:

1. Can long-term team performance be predicted on the basis of compositional characteristics?
2. Can long-term team performance be predicted on the basis of a short-format laboratory exercise?
3. What dynamic measurements are most effective at predicting long-term team performance?

Results are presented from a longitudinal study of graduate student design engineering teams working on innovative product development projects. Significant challenges were encountered in creating a compositional model for long-term innovative performance that appeared valid across the multiple years of analysis. Similarly, the short-format laboratory exercise proved to be somewhat unreliable in that it was not associated with any observed individual or group-level traits. However, significant opportunities were uncovered for developing a team dynamics-based model that moderates the relationship between team composition and long-term performance. An explanation of these challenges, as well as various opportunities for direct measurement of team dynamics are explored in context.

2 Expanded Data Set: Five Year Study

Our pilot study was conducted with the international population of ME310 students during the 2009–2010 academic year, including 100 students and 15 globally-distributed teams in total (Kress et al. 2012). Due to the exploratory nature of the study, a broad data set was collected including standard ethnographic data and responses to four separate questionnaire-based intrinsic measures. Several promising correlations emerged at all three levels of analysis (composition, dynamics and performance); an expanded study was designed to confirm the validity of the result across contexts by incorporating measurements from the 2 years immediately preceding and following the year of the pilot study, or a total of 234 students and 53 teams. Due to limitations on availability of archival data, the expanded data set is analyzed at the "local-only" level (collaboration with global partner teams is not taken into account).

2.1 Cognitive Style Data (Wilde Type)

On the basis of the preliminary analysis, the Wilde Type indicator emerged as the strongest of the four measures collected in terms of apparent statistical reliability and correlations with team-level observations (Kress et al. 2012). Archival Wilde

Table 1 Expanded data set – five year study

Academic year	2007–08	2008–09	2009–10	2010–11	2011–12	Total
Students	33	35	97	29	40	**234**
Teams	11	9	15	9	9	**53**

Type data were available from the ME310 teaching team for the two academic years preceding the pilot study (2007–2008 & 2008–2009); new data were collected for the two academic years following the pilot study (2010–2011 & 2011–2012). The Wilde Type data are comprised of individual responses to a 20-item online questionnaire that are combined to yield four independent "cognitive mode" preference scores (Wilde 2008). The four modes are generally distinguished by the "dominant functions" Thinking, Feeling, Sensing and Intuition, as initially described by Carl Jung (Jung and Hull 1971). A summary of student and team involvement in the study by year is provided in Table 1.

2.2 *Project Difficulty Assessment*

Each team receives a unique project prompt from an external client; therefore, projects range in content and difficulty. Prompts were collected and scored on a four-factor scale to establish initial difficulty as a basis for comparison. Scoring was conducted by crowdsourcing consensus values for each project along each of the four factors (ambiguity, breadth of scope, technical complexity, and overall difficulty) using five-point Likert-type scales in an online survey. Each prompt was scored a minimum of 25 times by independent respondents and the group mean is taken as the consensus score. Adding up the four factors yields a total possible difficulty range of 5–20; bivariate correlation tests confirm that the four difficulty factors are strongly and significantly positively correlated, suggesting that the additive approach is appropriate (see Table 2).

The 5 20 range suggests that the most difficult project is approximately four times as difficult as the easiest; we refer to this parameter as the difficulty ratio (α). The raw score (P_i) can thereby be adjusted by means of a linear multiplicative formula with a fourfold score amplification at maximum and zero amplification at minimum (see Eq. 1). A total of 125 respondents were recruited via Amazon Mechanical Turk to perform the ratings on a third-party survey website. Two fake prompts were inserted as a validity check (one easy and one difficult).

Equation 1: Project Performance Difficulty Adjustment Formula

$$P_A = P_i\left(1 + \frac{D_i - D_{MIN}}{D_{MAX} - D_{MIN}}(\propto -1)\right) \propto = \frac{D_{MAX}}{D_{MIN}} \quad (1)$$

The raw performance score P_i is adjusted by the specific project difficulty D_i, the difficulty limits D_{MIN} & D_{MAX}, and the difficulty ratio α to obtain the final score P_A.

Table 2 Correlation between difficulty factors

$n = 11$	Breadth	Complexity	Overall
Ambiguity	$r = 0.767, p = 0.006$**	$r = 0.646, p = 0.032$*	$r = 0.723, p = 0.012$*
Breadth	–	$r = 0.804, p = 0.003$**	$r = 0.812, p = 0.002$**
Complexity	–	–	$r = 0.957, p < 0.001$**

*Result is significant at the level $p \leq 0.05$; **Result is significant at the level $p \leq 0.01$

2.3 Project Performance Assessments

Documentation, images and other media of the final project outcomes were collected and archived. Because of the extensive variation between project content and documentation layout, prior to assessment the project materials were summarized into a standardized format providing representative information (e.g. problem statement, written summary, images, and design requirements) in a few pages at most. Using these summaries, a team of two graders trained in agreement assessed the projects in random order according to a four-factor framework as follows:

- **Usability** – the extent to which the proposed solution would be functionally useful by its intended target group were it to be released as an actual product.
- **Feasibility** – the extent to which the proposed solution is capable of being implemented as a real product with current technology, and that the team has demonstrated such.
- **Desirability** – the extent to which the product would be considered desirable by its target group, when considered in the real market context.
- **Originality** – the extent to which the product represents a unique insight or implementation, when considered in the real market context.

Each project receives a score for each of these four factors, with scores given according to a simple three-point rating scheme where 0 represents "Definitely Not," 1 represents "Arguably" and 2 represents "Definitely" (e.g. a 0 for Desirability indicates "Definitely Not Desirable"). The narrow rating scheme is designed to reduce the ambiguity inherent in the subjective assessment scheme. The four factors are designed such that a team will be penalized in scoring if it fails to clearly indicate a target user group or to demonstrate how the solution would be implemented as a real product. The graders' initial agreement was in excess of 75 %; after discussion they reached 94 % agreement. Final scores were computed as the mean of their two assessments. The scores for each factor are then added to yield the total raw assessment score (0–8 range). These raw scores are then adjusted for difficulty according to the multiplicative formula (see Eq. 1). The resulting adjusted performance score is on a 0–32 scale that is normalized as a percentage score. Project assessment and difficulty rating were conducted for the 53 projects in the data set spanning five academic years (Table 3).

In the 5-year sample, raw performance scores ranged from 1 to 8 and difficulty scores ranged from 9 to 15. The mean difficulty score was 11.7 with a standard deviation of 1.5 points. Though the rubric allows for a maximum-to-minimum

Table 3 Correlation between the performance factors

$n = 53$	Feasibility	Desirability	Originality
Usability	$r = 0.450, p = 0.001$**	$r = 0.716, p < 0.001$**	$r = 0.447, p = 0.001$**
Feasibility	–	$r = 0.170, p = 0.224$	$r = 0.367, p = 0.007$**
Desirability	–	–	$r = 0.393, p = 0.004$**

*Result is significant at the level $p \leq 0.05$; **Result is significant at the level $p \leq 0.01$

difficulty ratio of 4, the consensus ratings yielded a ratio of 1.67 suggesting that the most difficult project was only about 67 % more difficult than the easiest project. By both measures the variation in difficulty is fairly minimal. Applying the difficulty adjustment formula, adjusted performance scores were obtained and ranged from 2 to 24 of the total possible 0–32 range. Normalized as a percentage, the adjusted performances scores ranged from 6.1 to 74.8 % (see Fig. 1). There is a fairly continuous spread of scores in the 6–50 % range, with a slight gap just above the mean line and a large gap above the 50 % line; therefore the scores can roughly be classified as follows: Below Average, Above Average and Top Performers. The adjusted performance assessments were significantly positively correlated with the final grades the teams received independently from their teaching team ($r = 0.557$, $p = 0.002$, $n = 29$). No direct relationships were observed between performance and initial project difficulty.

Only four of the teams (or less than 8 %) score as Top Performers; these were the only teams to exceed 50 % on the performance scale. The majority of teams (33 teams, or > 60 %) are in the Below Average group with the remainder (16 teams, or 50 %) scoring just above average (see Table 4). The overall average performance score was surprisingly low at 29 % of the maximum. It is apparent that the four top performers are outliers of the otherwise continuous and nearly-linear distribution in adjusted performance scores. Therefore, there is an indication that something qualitatively different is happening within the top performing teams and so they ought to be considered independently in the analysis. We believe that this is evidence of a "dynamic transition" to a highly functional state that only a minority of teams appear to achieve naturally. Most teams do not undergo this transition and therefore are limited to the lower half of the performance spectrum.

The stability of performance scores over the 5 years of the expanded study, as well as the qualitatively continuous distribution of scores taken together (excluding top performers) offers some support for the reliability and objectivity of the performance-rating rubric. There was no significant change observed in mean performance year over year, though there did appear to be an expansion in the range of scores in the penultimate year followed by a contraction and less than significant decrease in the final year of analysis (see Fig. 2). A t-test confirms that the null hypothesis cannot be rejected, therefore these qualitative variations are not significant and the performance scores are comparable across years.

None of the correlations between Wilde mode composition and team performance that were observed in the pilot study were entirely consistent over the 5-year span (see Table 5). This result can be interpreted in two ways: either that the

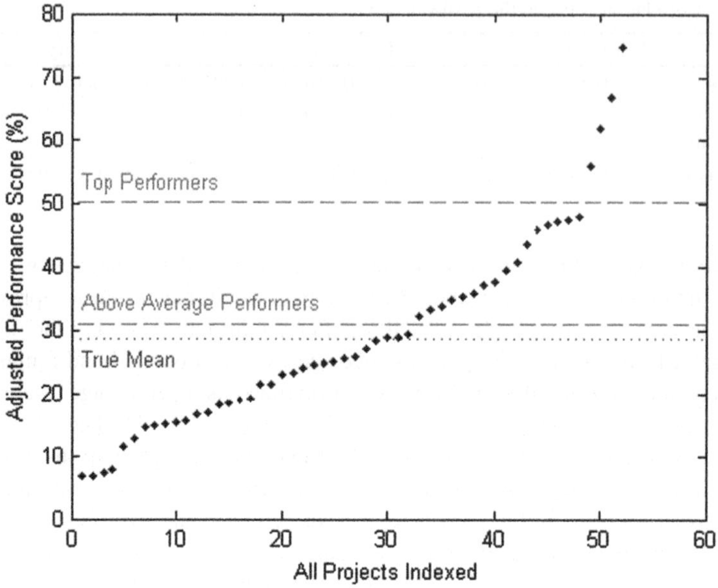

Fig. 1 All performance scores (adjusted for difficulty and sorted)

Table 4 Performance scores by group

Group	Approx. range	Actual range	No. of teams in group
Below average	0–30 %	6.09–29.31 %	33
Above average	30–50 %	32.22–47.84 %	16
Top performers	50–100 %	55.81–74.81 %	4

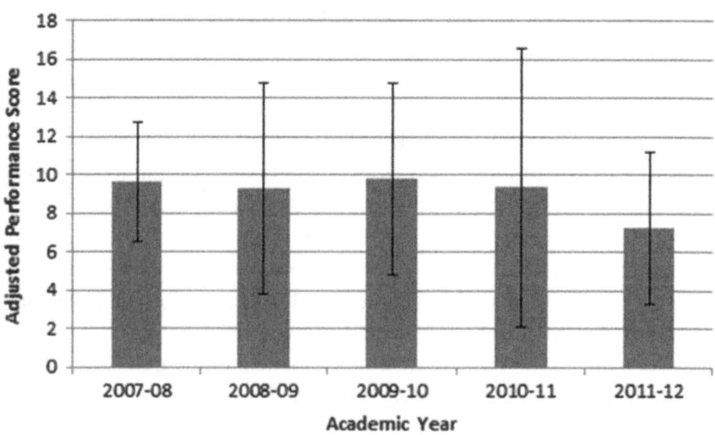

Fig. 2 Mean performance by rear

Table 5 Wilde mode correlations with performance by year

Year/mode	Ext. intuition	Ext. sensing	Ext. thinking	Ext. feeling
2007–08	Trends negative	Trends negative	Trends negative	Trends negative
2008–09	No trend	No trend	No trend	No trend
2009–10	No trend	Trends positive	Trends negative	Positive
2010–11	No trend	No trend	Trends negative	No trend
2011–12	Negative	Trends positive	Positive	No trend

correlations observed in the pilot study were merely statistical artifacts, and therefore are not evidenced in the broader sample; or that there were considerable year-to-year variations in the team context that introduce too much noise to draw a meaningful comparison. It is a fact that there were substantial changes to the course structure, curriculum and personnel during each of these 5 years, though we had no explicit measures of these factors and were unable to account for them in the analysis. It is likely that these year-by-year variations led to the development of different working environments for the teams, thereby favoring different compositional characteristics. The volatility of these results to changing contexts – and the overall low performance mean – underscores the need for an improved dynamic model for observing, explaining and influencing team effectiveness.

3 The SDTE Approach

The scope of the expanded study, in excess of 50 unique projects spanning 5 years, particularly highlights the need for a standardized measurement of team function that does not rely on after-the-fact subjective assessments of the project outcome. Ideally, this measurement would require a minimum of time, effort and equipment on behalf of the students and researchers (or managers or employees, or whatever the case may be). This was the inspiration to develop a short-format, in-laboratory exercise to capture individual and team-level reframing behaviors as a proxy for long-term innovative performance. In total, 15 trials were conducted with the final version of the exercise as part of a pilot evaluation and validation. Further experimental trials with new product development teams are ongoing at the Luleå Institute of Technology (Division of Innovation and Design).

3.1 Development of the Exercise

The instrument is an episodic case study that requires both individual and group decision-making in the form of choice structuring. The subject of the case study was an "urban public share bicycle" such that subjects are asked to design a bicycle suitable for an urban environment and shared by the general public. This topic was chosen because it struck a balance of familiarity; most subjects are familiar with bicycles, although perhaps not in this context or in great depth. The case study involves a short text introduction describing the scenario and presenting the subject with 10 initial design choices. The subject is asked to rank those choices from 1 (most important) to 10 (least important). This process of choice structuring within a case study as a measure of team decision-making has been applied in prior research (Artiz and Walker 2009). New information is introduced in a series of three rounds, providing a number of opportunities for measuring choice restructuring (reframing). We theorized that the multi-level measurement of reframing behavior would give insight into the health of the team dynamic as well as provide evidence of underlying interpersonal intrinsic differences. A complete discussion of the development and application of the STDE instrument can be found in the previous installment in this series (Kress and Schar 2012).

3.2 SDTE and Social Sensitivity

Prior research has suggested that the only intrinsic compositional characteristic with a reliably positive impact on team function is the presence of individual-level "social sensitivity" (Woolley et al. 2010). Social sensitivity is measured by the "Reading the Mind in the Eyes" (RME) test which assesses an individual's ability to interpret the emotions of others by viewing a series of photographs of strangers' faces paired with single-word answers in a multiple choice format (see Fig. 3). There is a correct answer for each photograph, and the resulting score is simply the individual's total number of correct answers (Baron-Cohen et al. 2001). Elsewhere this test has been used as a direct measure of subject empathy.

As a known correlate of project performance and a direct moderator of interpersonal interactions, we theorized that individual differences in social sensitivity would correspond with variations in reframing. For example, heightened social sensitivity could facilitate consensus-finding or empathy with the hypothetical end-user. However, no relationships were observed between social sensitivity and reframing behavior at either the individual or team levels. This suggests either that reframing behavior is not reliably influenced by underlying social sensitivity, or that the instrument is not adequately capturing interpersonal variations in predisposition toward reframing. In either case, it means that the SDTE cannot be used to understand the means in which social sensitivity positively influences team performance.

3.3 SDTE and Cognitive Style

The Wilde Type measure of cognitive style is theoretically a collection of individual-level traits differentiated by the four dominant intrinsic functions.

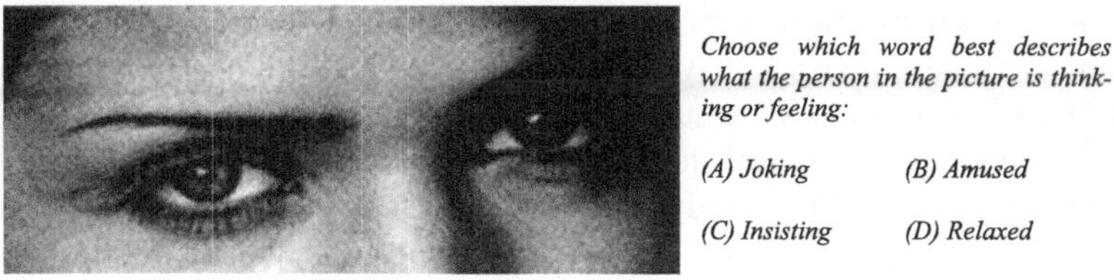

Fig. 3 A sample item from the "Reading the Mind in the Eyes" test

Because they are traits, we expect that they are normally distributed in the population and change either not at all or very slowly throughout a person's lifetime. Because the modes are differentiated by function, we expect that they at least partially describe the functional contribution that an individual team member could be expected to make to the team's progress. Therefore, correlations between individual-level reframing and Wilde mode preferences could indicate that the tendency to reframe is also a stable individual trait with reliable impacts on teamwork. Unfortunately, no such relationships were observed; reframing behavior did not appear to be reliably affected by underlying Wilde cognitive mode compositional characteristics. These results suggest either that the output of the exercise is too noisy to discern compositional effects, or that reframing behavior is not reliably influenced by intrinsic function.

3.4 Summary of the SDTE Approach

In conclusion, we were unable to show that performance on the laboratory exercise corresponded with known compositional measures that are important for teamwork (i.e. social sensitivity and Wilde mode characteristics). This poses a threat to the usefulness of the SDTE as it does not appear to capture stable trait-level variations; rather, the data appear unreliable and heavily influenced by the immediate context and team dynamics conditions. Therefore we cannot claim that reframing behavior as captured by SDTE is a meaningful reflection of team compositional characteristics nor that it is likely to be representative of long-term team performance. Contrary to our intent, it appears that performance on the instrument is highly context-dependent and most strongly modified by team dynamics factors that were not accounted for. As such the SDTE was excised from the remainder of the study and more promising dynamics-level measurements were pursued.

4 TeamSense: Improved Dynamics Measurement

Our prior research encountered serious challenges in attempting to develop a compositional model of teamwork that was valid across contexts. Furthermore, our attempts to develop a standardized exercise to provide a quick quantitative measure of team effectiveness (via the proxy of reframing behavior) were similarly frustrated. These experiences have led us away from the traditional observational or manually-coded dynamics measures in favor of direct measures that provide quantitative data coupled with real-time computer assisted analysis. We see significant opportunities to augment team dynamics research in particular with the increasing availably and ubiquity of low-cost sensing systems and further development of rapid prototyping platforms for hardware and software.

4.1 Need for Improved Dynamics Measurements

The difficulty in creating a compositional or longitudinal performance-based model of team performance – particularly one that is valid across changing team contexts – has encouraged us to pursue direct measures of those team dynamics characteristics that are indicative of both short and long-term team effectiveness. Managers and teachers in project-based courses rarely have the time or opportunity to observe teamwork in progress, and so may miss critical dynamics cues. Additionally, they may simply be unaware of what those cues are or how to address them. We believe that real-time, in situ team dynamics monitoring could have substantial benefit to team performance and learning, both through direct feedback to the team and through mediated feedback (e.g. teacher, coach or manager). We propose a distributed network of unobtrusive, "ambient" sensors to measure team function in real time and a series of prototypes to present this information in a useful and impactful format for team members and those overseeing them.

4.2 Background: Embedded Sensing

Embedded sensing utilizes low-cost computation platforms, such as programmable microcontrollers, to continuously collect and process data from attached sensor modules. For example, a microprocessor connected to an accelerometer and a small battery can be placed in the team environment, and programmed to send events when a teams physical activity exceeds some define threshold. While embedded systems are cost-effective ways to sense the environment, they historically have required specialized technical background in both electronics hardware and software. The use of novice-oriented microcontroller based platforms, such as Arduino, have shown increasing adoption with non-technical users and suggest new ways to rapidly prototype platforms specific to human activity and team dynamics measurement.

4.3 Guiding Questions and Research Approach

How might we facilitate better teamwork through real-time monitoring and feedback of team dynamics factors in the workplace? Can we create quantitative metrics for team dynamics that are reliable and meaningful predictors of team performance? Can we create a platform that allows even non-technical users to create and analyze their own sensing experiments? To answer these questions we propose the "TeamSense System" – a modular platform for precision real-time data capture, analysis and feedback. We envision that a combination of (i) unobtrusive sensing hardware in the collaboration environment, (ii) software analysis tools for detecting patterns of team activity and, (iii) dynamic feedback mechanisms for behavioral intervention, may radically accelerate the collaborative design process. The general proposed of research approach is as follows:

(i) **Identify metrics and develop a variety of hardware prototypes.** To directly sense different measures of Sensors will be selected based on their qualities of minimal obtrusiveness, flexibility, modularity and low-cost. A particular emphasis is placed on simple "low fidelity" or ambient sensors that can be easily deployed in the field with off-the shelf technology, and easily converted to low-dimensional quantitative metrics.

(ii) **Data capture and feedback in the field.** We have completed a pilot study with a single type of analog sensor (piezoelectric vibration sensor) with an initial sample of two ME310 student design teams. An expanded study will observe (a) a larger number of student design teams at Stanford (both ME310 and d.school teams) and (b) global design teams at partner sites such as the Hasso Plattner Institute. Here we will explore different methods of feedback of the data to the teams and those overseeing them, and measure the effects of these interventions.

(iii) **Data analysis and evaluation.** Data streams from sensors will be analyzed both in real-time and offline for indicators of salient team dynamics factors. Depending on the sensor, the analysis may range from simple measures, such as frequency of events and average clustering, to more sophisticated machine learning techniques. We can then compare data streams with existing models of team performance, including leveraging techniques from prior work on linking cognitive affinities with team dynamics and performance.

4.4 Prototype Development

In order to demonstrate the feasibility of the concept we constructed and tested two sample prototype TeamSense units at Stanford in Spring Quarter 2012 (see Fig. 4). These sensors aim at capturing the amount of physical activity that takes places during teamwork. Each unit was designed to be mounted on the underside of a team's table in the shared workspace. It registers physical activity in real time by

Fig. 4 Two prototype sensing units for mounting beneath a table, showing power connection on one unit and wireless transceiver (connected to PC)

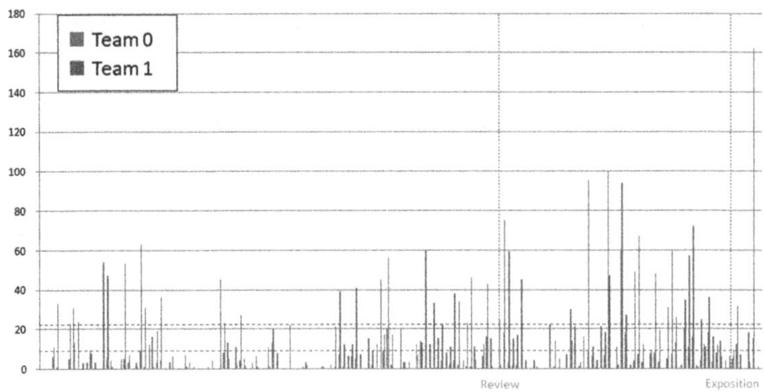

Fig. 5 Data collected from two different ME310 teams over the same time 6-week period shows quantitative and qualitative differences in team activity

means of a piezoelectric vibration-sensing contact microphone, and transmits this information wirelessly to a base station PC along with the Team ID. This allows for real-time monitoring of team activity as well as data-logging for asynchronous team observation. The platform was designed to be modularly expandable to any number of sensing units operating in the field.

Data was collected for two ME310 teams over a 6-week period as part of the pilot testing of our prototype units. The data demonstrate the feasibility of measuring team activity with unobtrusive sensors, and clearly reveals quantitative and qualitative differences in the pattern of activity of two ME310 teams (see Fig. 5). However, testing with these initial prototypes uncovered the need for additional calibration steps and improved signal processing for future devices. We aim to expand on this approach to explore a variety of different sensing techniques and analysis, as well as to increase the sample size of teams and introduce additional pathways for validation. We will also continue to pursue these dynamic measurements as the basis for a model of team function that can meaningfully connect compositional characteristics and longitudinal performance.

In conclusion, our attempts to observe consistent compositional effects in the expanded study, as well as our attempts to develop an exercise to measure team effectiveness, were both frustrated by the heavily context-specific nature of teamwork. This research has identified the need for improved team dynamics measures from multiple perspectives. Our aim is to develop a means of team measurement that is meaningful across different contexts by capturing those aspects which are most critical for successful team function. Our work to date, as well as emerging research in the field indicates that direct quantitative measurement of the team dynamic is the most promising avenue for development (Jung et al. 2012; Jung et al. 2011; Sosa 1999; Woolley et al. 2010). Though the technical barriers to this type of measurement are rapidly falling, there remains considerable work to understand what measurements are the most meaningful, how they are best obtained (so as not to hinder team function) and how they can most effectively be used to improve team performance either through intervention by a mediator or direct feedback from an automated system.

References

Artiz J, Walker RC (2009) Cognitive organization and identity maintenance in multicultural teams: a discourse analysis of decision-making meetings. J Bus Commun. doi:10.1177/0021943609340669, http://job.sagepub.com/content/early/2009/07/16/0021943609340669

Baron-Cohen S, Wheelwright S, Hill J, Raste Y, Plumb I (2001) The 'Reading the Mind in the Eyes' test revised version: a study with normal adults, and adults with Asperger syndrome or high-functioning autism. J Child Psychol Psychiatry 42(2):241–251, doi:10.1111/1469-7610.00715

Jung CG, Hull RFC (1971) Psychological types. Bollingen series XX (trans: Baynes HG). Princeton University Press

Jung MF, Larry JL, James JG, Pamela Hinds, Ralf Steinert, Stanford University. Department. of Mechanical Engineering (2011) Engineering team performance and emotion affective interaction dynamics as indicators of design team performance. http://purl.stanford.edu/th996ft5752

Jung M, Chong J, Leifer L (2012) Group hedonic balance and pair programming performance: affective interaction dynamics as indicators of performance. In: Proceedings of the 2012 ACM annual conference on human factors in computing systems, pp 829–838. CHI'12. ACM, New York. doi:10.1145/2208516.2208523. http://doi.acm.org/10.1145/2208516.2208523

Kress G, Schar M (2011) Initial conditions: the structure and composition of effective design teams. In: Proceedings of the 18th international conference on engineering design (ICED11), vol 7, pp 353–361

Kress GL, Schar M (2012) Applied teamology: the impact of cognitive style diversity on problem reframing and product redesign within design teams. In: Hasso P, Christoph M, Larry L (eds) Design thinking research, Understanding innovation. Springer, Berlin/Heidelberg, pp 127–149, http://www.springerlink.com/content/g658277881645046/abstract/

Kress G, Steinert M, Price T (2012) Cognition as a measure of team diversity. In: Interdisciplinary engineering design education conference (IEDEC), 2nd edn., pp 67–72 doi:10.1109/IEDEC.2012.6186925

Skogstad P, Steinert M, Gumerlock K, Leifer L (2009) We need a universal design project outcome performance measurement metric: a discussion based on empirical research. In: Proceedings of the 17th international conference on engineering design (ICED'09), vol 6, pp 473–484

Sosa R (1999) Modelling creative design through conversation analysis. In: Proceedings of the 3rd conference on creativity & cognition, C&C'99. ACM, New York, pp 182–183. doi:10.1145/317561.317592. http://doi.acm.org/10.1145/317561.317592

Tang JC (1991) Findings from observational studies of collaborative work. Int J Man–Mach Stud 34(2):143–160, doi:10.1016/0020-7373(91)90039-A

Tang JC, Leifer LJ (1991) An observational methodology for studying group design activity. Res Eng Des 2(4):209–219, doi:10.1007/BF01579218

Wilde DJ (2008) Teamology: the construction and organization of effective teams. Springer, New York

Woolley AW, Chabris CF, Pentland A, Hashmi N, Malone TW (2010) Evidence for a collective intelligence factor in the performance of human groups. Science 330(6004):686–688, doi:10.1126/science.1193147

Early and Repeated Exposure to Examples Improves Creative Work

Chinmay Kulkarni, Steven P. Dow, and Scott R Klemmer

Abstract This article presents the results of an online creativity experiment (N = 81) that examines the effect of example timing on creative output. In the between-subjects experiment, participants drew animals to inhabit an alien Earth-like planet while being exposed to examples early, late, or repeatedly during the experiment. We find that exposure to examples increases conformity. Early exposure to examples improves creativity (measured by the number of common and novel features in drawings, and subjective ratings by independent raters). Repeated exposure to examples interspersed with prototyping leads to even better results. However, late exposure to examples increases conformity, but does not improve creativity.

1 Introduction

Examples are considered "a cornerstone of creative practice" (Herring et al. 2009). Leveraging examples of prior work is an established technique in design (Buxton and Buxton 2007), and many design programs encourage students to use examples of existing designs (Schön 1985). However, the strategies employed by designers to seek and use examples is largely ad-hoc (Newman and Landay 2000).

Frequently, these strategies differ in timing – in an informal survey we conducted among designers around Stanford University, one respondent described inspirational examples as "huge parts of my initial steps. I need to know as much as I can about the topic before I feel comfortable moving forward." In contrast, another said that "I don't do this [look at examples] at the very beginning because

C. Kulkarni (✉) • S.R. Klemmer
Stanford University HCI Group, Stanford, CA 94305-9035, USA
e-mail: chinmay@cs.stanford.edu; srk@cs.stanford.edu

S.P. Dow
Carnegie Mellon, HCI Institute, Pittsburgh, PA 15213, USA
e-mail: spdow@cs.cmu.edu

H. Plattner et al. (eds.), *Design Thinking Research*, Understanding Innovation,
DOI 10.1007/978-3-319-01303-9_4,

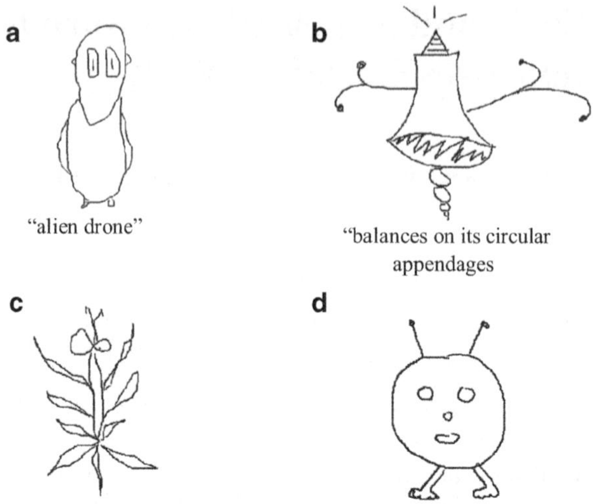

Fig. 1 Sampling of drawings created in our experiment, with excerpts of participant-provided descriptions

(a) "alien drone"

(b) "balances on its circular appendages

(c) "bug-like alien"

(d) "This round furry creature gets around gets around either by walking on his retractable legs or rolling across surfaces."

it gets your mind stuck in one way of thinking." This fear of conformity was echoed by other participants, and one went on to say that he looked for inspiration only when "facing a creative block."

These different strategies suggest that examples may modify the creative process differently depending on the point in the design process at which they are presented. This leads to the practical question: what are the tradeoffs of looking at examples earlier or later in the design process? Furthermore, even if there is an "ideal" time to view examples, some designers feel ubiquitous information access and their own "thirst for knowledge" bombards them constantly with examples (Herring et al. 2009). How does this repeated exposure to examples affect the creative process?

This article presents the results of an online creativity experiment we conducted on Amazon Mechanical Turk. Participants in the experiment generated drawings of alien creatures as a creative task. The pervasive use of sketches to develop and communicate conceptual designs in the creative fields (Suwa and Tversky 1997), and the use of similar tasks in prior work Ward (1994) inspired the choice of the drawing task. Focusing on drawings of alien figures makes this task readily accessible to non-designers (see Fig. 1 for a sampling of drawings created by participants).

Participants were randomly assigned to one of four conditions: examples early, examples late, examples early and late, or a control condition without examples. This study's creativity measures were the number of uncommon and novel features in the drawings and Likert-scale ratings by condition-blind raters. Conformity was measured by the number of critical features (features that were directly copied from examples).

This paper's experimental results suggest that while exposure to examples increases conformity, such exposure early in the creative process improves the creativity in the output, while later exposure provides no such benefit. Furthermore, exposure to examples followed by prototyping and subsequent re-exposure to the same examples improved creative output even more. This finding may allay some fears of example bombardment. Lastly, in our experiment, participants exposed to examples created fewer drawings, so these example driven quality improvements may come at the cost of a lower quantity of creative work.

2 Related Work

2.1 *Examples*

Bringing existing solutions to mind is crucial for creative generation (Smith et al. 1993). The Structured Imagination theory by Ward (1994) describes creativity as a multistep process: in the recall step, people bring to mind existing solutions and constructs. Then, in the modification step, these constructs are altered in novel ways. Similar analogical processes are found in other areas of cognition such as analysis and learning (Gentner and Colhoun 2010).

Designers often incorporate features from examples directly into their work (Marsh and Bower 1993, "inadvertent plagiarism"); but examples also "ultimately alter the nature of the creative product" in more subtle ways (Marsh et al. 1996). Lee et al. (2010) found that designing with examples generally improves the quality of creative work. These findings have also led to tools for discovering, storing and retrieving examples (Kerne et al. 2008; Ritchie et al. 2011).

The current work is an extension of Marsh et al. (1996) (itself an extension of Smith et al. 1993), so we describe Marsh et al.'s experiment in more detail. In their experiment, participants generated drawings of non-Earth-like creatures to inhabit an alien planet similar to Earth. In the example conditions, experimenters provided participants example drawings of aliens at the start of the experiment. Example drawings all had certain attributes, or critical features, in common – four legs, antennae and a tail. The proportion of these critical features incorporated into participants' own drawings was used as a measure of conformity. The proportion of other, noncritical, features was used as a measure of creativity. These non-critical features were classified as either novel (not commonly found on animals, such as speakers or propellers), uncommon (such as a pouch or tentacles), or common (such as a nose, mouth or two legs).

Participants exposed to examples incorporated more critical features in their drawings, but not at the expense of novel and uncommon features. Instead, their drawings contained fewer common features. This suggests that while examples increase conformity by increasing activation of critical features, they do not block retrieval of original ideas (such as novel and uncommon features).

We use Marsh et al.'s feature-based evaluation metric, and extend their work by examining how the example timing affects creative output. In addition, we study the effects of repeated exposure to examples in the creative process.

2.2 Research Methods

Our experiment uses a task (drawing sketches of alien figures) that has previously been employed to study creativity in a context of no prior training (Marsh and Bower 1993; Ward 1994). Drawing tasks have also been demonstrated to be appropriate for online experiments (Yu and Nickerson 2011).

This experiment was run on Amazon Mechanical Turk (www.mturk.com), a web-based crowd-sourcing platform. This platform has been used for experiments on affect and creativity (Lewis et al. 2011). Mechanical Turk workers have also been employed to provide perception responses (Heer and Bostock 2010), objective labels (Deng et al. 2009; Snow et al. 2008), and subjective ratings (Dow et al. 2011).

3 Experiment

Our experiment had two goals. First, we wanted to see if exposure to examples at the start of a creative process leads to a different quality of creative output in contrast to exposure when the creative process is underway. Second, we wanted to investigate the role of repeated exposure to examples.

Our initial hypothesis was that exposure to examples later in the creative process would have the same creative benefits but lower conformity than exposure at the start. This hypothesis was motivated by Weisberg (1999), who observed that creative failures are more often explained by the absence of relevant information than the presence of irrelevant information.

Furthermore, the presence of one's own ideas would inhibit the adoption of sub-optimal ideas from late exposure to examples (mirroring the intuitions of some designers).

In the case of repeated exposure, the activation account would predict that, similar to showing more examples at once, this would result in greater degree of conformity due to higher activation of features present in examples.

3.1 Participants

We solicited US-resident participants on Mechanical Turk with a compensation of US$1.00. 81 participants responded (27 male, 54 female; median age 34). All participants reported a high-school diploma or a higher degree. This between-subjects experiment randomly assigned participant to one of four conditions.

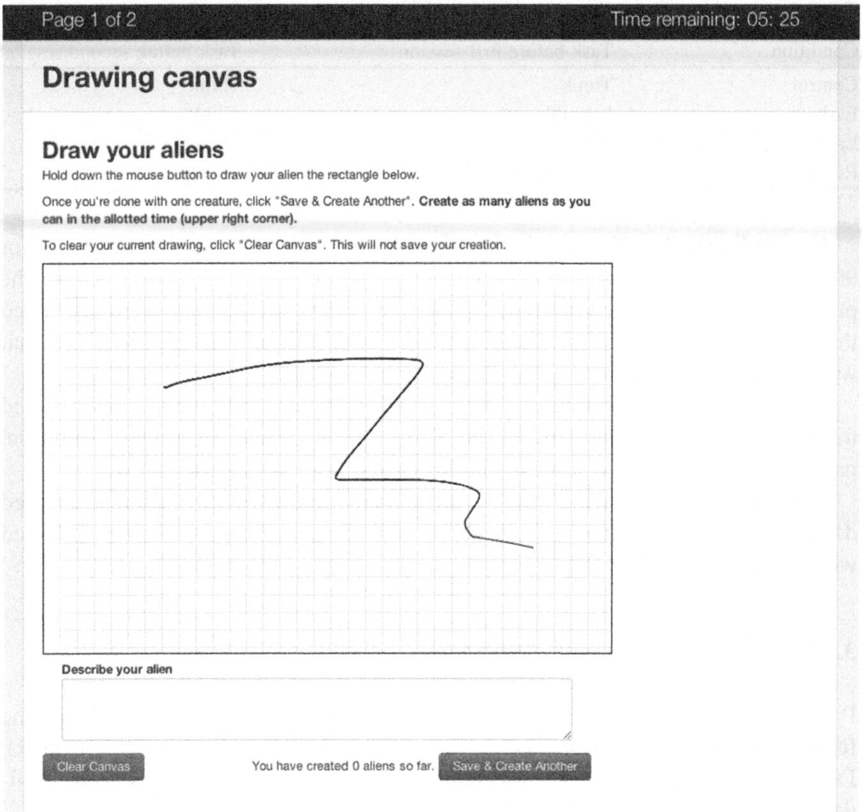

Fig. 2 Drawing canvas with time remaining (*top right*), and an option to clear the canvas (*bottom left, red*)

3.2 Procedure

The experiment comprised two drawing sessions, each lasting 7 min. Participants were asked to create as many drawings as they could during the drawing session. To encourage this (and discourage participants from spending time perfecting only a few drawings), the experimental platform included a clear-canvas tool but no line-eraser tool (Fig. 2).

Each session was preceded by a condition-specific task in which partic pants were either exposed to examples, or asked to think about the aliens they planned to draw in the next session (Table 1).

At the start of the experiment, all participants saw a Web page with instructions adapted from Marsh et al. (1996) to account for two drawing sessions and a break (see hci.stanford.edu/example/aliens for actual prompts used).

Table 1 Experimental conditions

Condition	Task before first session	Task before second session
Control	Think	Think
Early	Examples	Think
Late	Think	Examples
Repeated	Examples	Examples

For the **Example** task, participants were shown three example alien drawings for 90 s (see Fig. 4). We used drawings from (Marsh et al. 1996), p. 672. Using the prompt of Marsh et al. (1996) (and Smith et al. 1993), participants were instructed that examples were only shown to help them create their original creations, and that we did not want them to copy the examples in any aspect.

For the Think task, participants were asked to "think about aliens" they planned to draw in the next session for 90 s. In the Repeated Examples condition, participants saw the same three examples before both drawing sessions.

After the second drawing session, participants filled out a survey that covered demographics, artistic interest and ability and the thought-process they followed while drawing.

3.2.1 Labeling Features in Drawings

Participants generated a total of 543 drawings. Each drawing was labelled with the features it incorporated from the feature set of Marsh et al. (1996) (Appendix). Drawings were annotated on Mechanical Turk, since the features were well-defined.

All workers were US resident and at least 18 years of age, and were compensated US$0.50 for the task. Workers who participated in the experiment were disallowed from the annotation task (and vice-versa). All annotators were blind to experimental condition.

Workers were trained using a drawing from a pilot participant (Fig. 3). Then, each worker annotated a set of seven randomly assigned drawings. Workers also rated how creative they found the drawing on a seven-point Likert scale (each annotator saw at least one drawing from each condition). Lastly, annotators could flag offensive (or non-alien) drawings. Upon review, 34 flagged drawings were discarded by the authors. Each drawing was annotated by two workers. Disputes in annotation were resolved by the authors (Fig. 4) (Table 2).

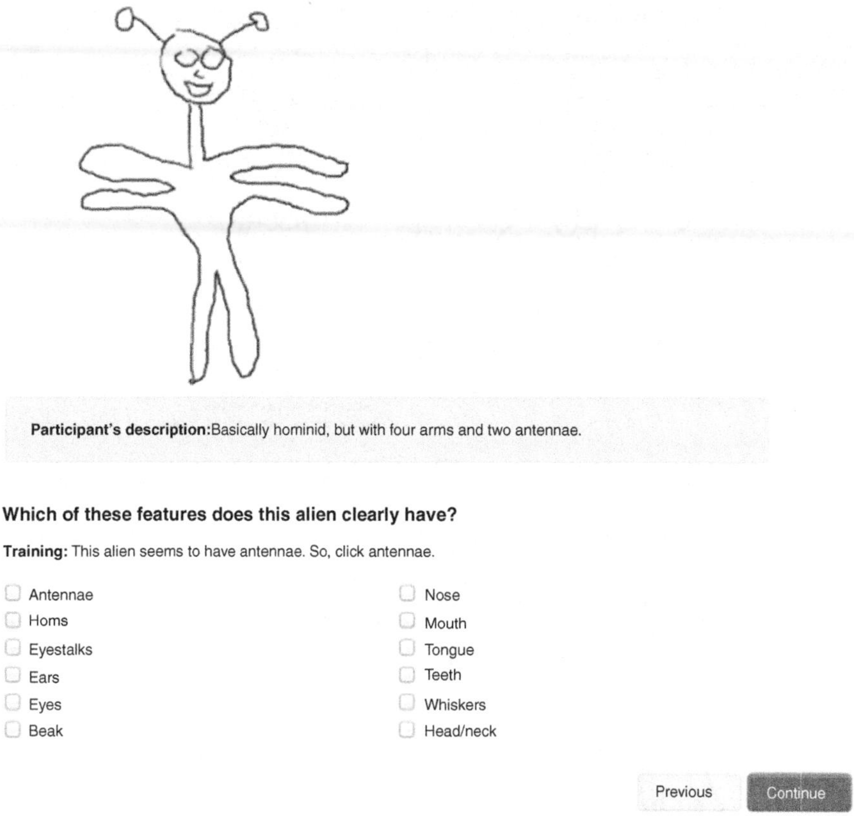

Fig. 3 Training interface for annotators. The training interface shows what features to label ("click *antennae*"). The actual annotation is performed on an identical list of features

4 Results

We analyze data using a mixed-effects linear model. Since participants drew multiple drawings per drawing session, unless noted, we consider the participant as a random effect with a fixed intercept; and experimental condition, drawing session (first or second), and an interaction term as fixed effects in all our analysis below. Reported p-values are from a Monte-Carlo (MCMC) simulation (Baayen et al. 2008).

4.1 Examples Increase Conformity

Following Smith et al. (1993), conformity was measured as the number of critical features incorporated per drawing. Without controlling for the drawing session, examples shown at the start of the experiment increased the number of critical

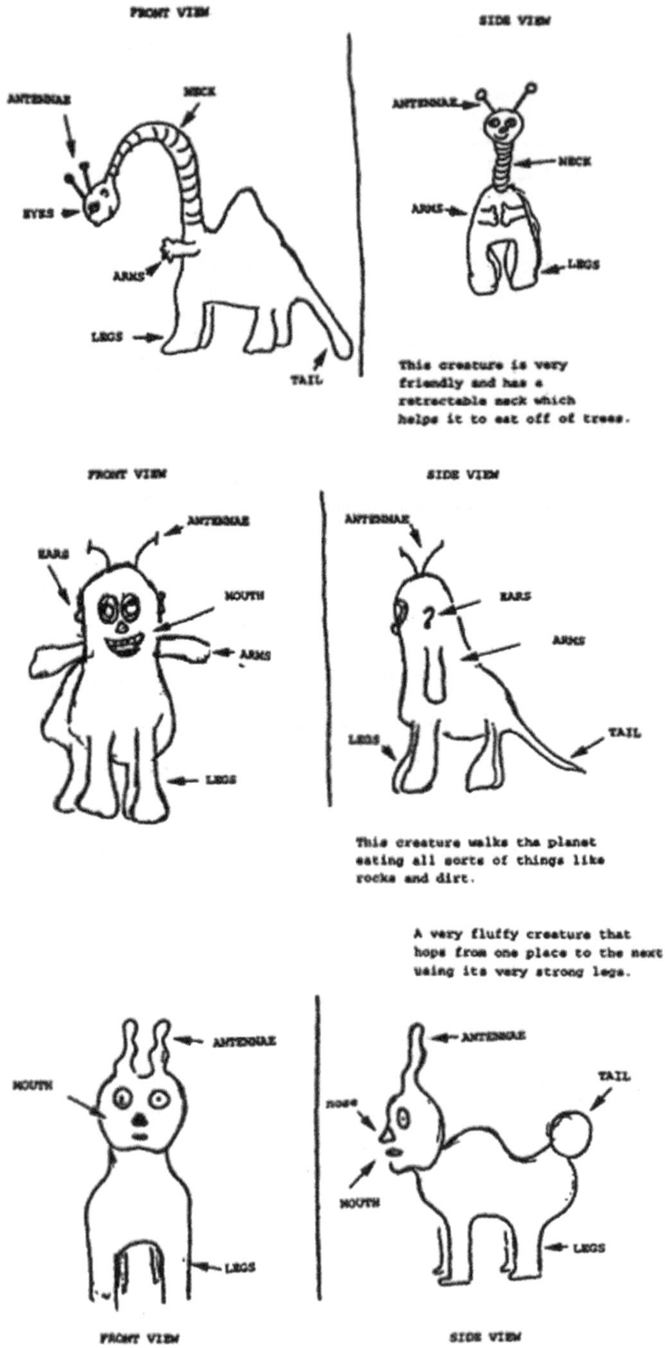

Fig. 4 Example drawings provided to participants. All examples contain critical features – four legs, antennae, and a tail

Table 2 Table of means. Means that differed from control at $p < 0:05$ are bold, those marginally significant ($p < 0:1$) are in italics (p-values from the post-hoc analysis using mixed models, see Sect. 5)

Condition	Critical	Common	Uncommon	Novel	Total	Drawing per session	Likert rating
Control	0.39	4.21	0.95	0.47	6.03	4.00	3.71
Early	**0.57**	3.91	*1.15*	0.40	6.04	**3.00**	**4.10**
Late	**0.52**	*3.82*	0.78	0.45	*5.57*	**3.68**	3.43
Repeated	**0.64**	4.20	**1.21**	0.54	6.60	**3.00**	**4.22**

features that were incorporated into drawings ($t(507) = 2:06$, $p < 0:05$), consistent with results from (Smith et al. 1993; Marsh et al. 1996). Participants in the Late Examples condition show higher conformity in the second drawing session (i.e. post-exposure) [$t(419) = 1:83$, $p = 0:07$].

4.2 *Early Exposure Increases Uncommon Features*

The number of uncommon features per drawing increased in the Early Examples condition ($t(419) = 1:61$; $p = 0:06$), and in the Repeated Examples condition ($t(419) = 1:72$; $p < 0:05$), but not in the Late Examples condition ($t(419) = 0:45$; $p = 0:649$) (Fig. 5). The number of novel features did not vary significantly across condition. Participants in the Late exposure condition created drawings with marginally fewer common features($t(419) = -1:33$; $p = 0:09$) and fewer total number of features ($t(419) = -1:30$; $p = 0:09$).

4.3 *Early and Repeated Exposure Leads to Higher Subjective Ratings*

Annotators rated drawings in the Early Examples and the Repeated Examples conditions higher ($t = 2:24$; $p < 0:05$ and $t = 2:65$; $p < 0:01$, respectively). Intra-class correlation amongst raters (average, random raters) was 0:54 ($F(508;508) = 2:2$; $p < 0:001$).

4.4 *Examples Reduce Number of Drawings*

Unlike Marsh et al. (1996), participants created fewer drawings per session in all example conditions[1] [Early: $t(149) = -2:50$; $p < 0:05$; Late: $t(149) = -2:14$,

[1] Since the number of drawings is not a repeated measure, analysis uses a fixed-effects model with interaction, the experimental condition and the type of session being independent variables.

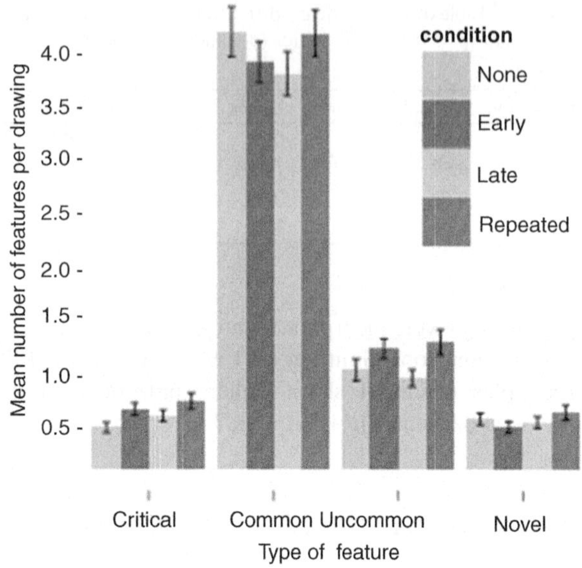

Fig. 5 Participants in early and repeated exposure conditions included more uncommon features compared to late exposure/control conditions

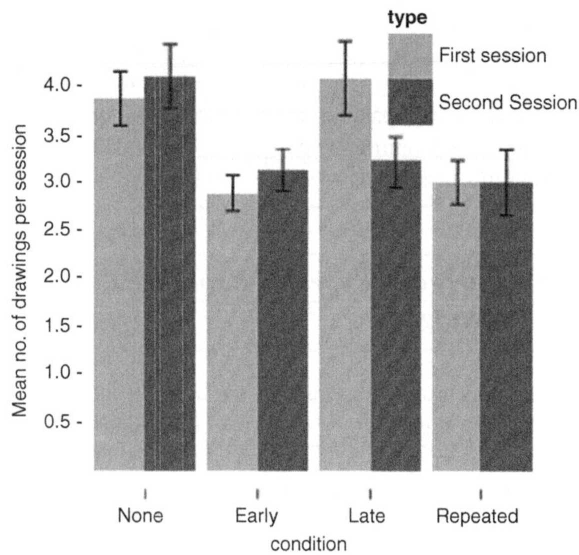

Fig. 6 Participants drew fewer drawings when examples were shown (in the late examples condition, participants drew fewer drawings in the second session)

$p < 0{:}05$, Repeated: $t(149) = -2{:}63$, $p < 0{:}05$] (Fig. 6). Participants in the Late Examples condition created fewer drawings after exposure to examples ($\mu_{be f ore} = 4{:}10$, $\mu_{a f ter} = 3{:}13$, $t(149) = 1{:}91$, $p_{interaction} < 0{:}05$).

5 Discussion

5.1 *Example Timing Affects Creative Output*

These results suggest that exposure to examples at any time increases conformity. However, early exposure increases the number of uncommon features and subjective ratings of creativity, while late exposure provides no such benefits. This runs counter to both our initial hypothesis and the intuitions of many designers who delay looking at examples in an effort to reduce fixation and think "out of the box" (Jansson and Smith 1991).

One possible explanation for these effects is that early exposure to examples aids the designer in understanding the scope of acceptable solutions to a problem, and helps form an initial representation of the creative concept (Heit 1992). Prototyping results in subsequent abstraction and refinement of the initial representation (Lim et al. 2008). Without initial exposure to examples, the refined representation may differ widely from the one embodied in examples, which would make it harder to map concepts from the example to one's own representation. When exposure is only for a short duration (90s in our experiment), it is possible that only concepts with high enough activation, such as critical features in our experiment, are transferred (motivated by Boroditsky 2007).

Another counter-intuitive experimental result is that repeated exposure to the same examples led to higher creative quality. This may also be explained by a seeding-and-transfer account. Initial exposure to examples prevents the refined representation formed by prototyping from diverging greatly from the one embodied in the examples. This refined yet similar representation would then allow the designer to learn different concepts on re-exposure to the same example.

In essence, the crucial ingredient that allows repeated exposure to improve creativity might be the prototyping that occurs between exposures.

5.2 *Why Did Examples Yield Fewer Drawings?*

Examples play a dual role in design– first, they inspire different solutions and ways of thinking. Second, they help form expectations about what characteristics a solution needs to have (Herring et al. 2009). The decrease in the number of drawings created may be due to this second role. Seeing examples may have signaled a higher threshold for "acceptable" drawings, resulting in participants spending more time on each drawing, and creating fewer drawings overall.

Our data suggest that this expectation-setting role has a different behavior than the inspirational role. While the number of drawings created decreased nearly uniformly post-exposure, changes in creativity measures (uncommon features and subjective ratings) were non-uniform. Therefore, while examples may set

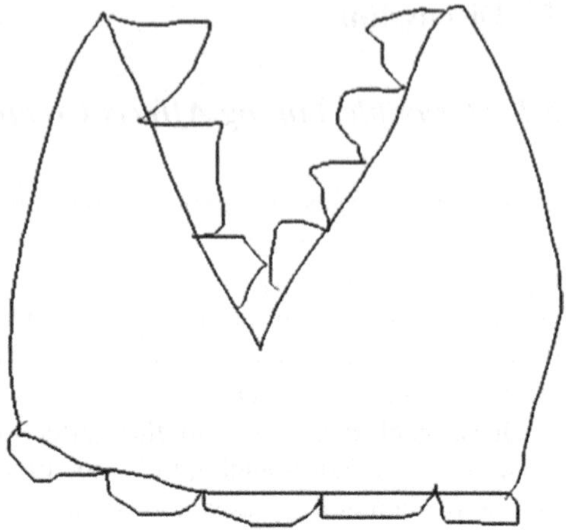

Fig. 7 Participant-provided description: "An ambush predator that does move very much, but lures prey into its mouth using scent to make them think there is food there. It only occasionally shifts using the pads on its bottom, which can also suck up nutrients from the ground or water for emergencies." Rated highly creative by our raters, this drawing has no novel or uncommon visual features, and uses a non-visual feature (scent)

expectations any time they are presented (including late in the design process), their inspirational value may be time-dependent.

5.3 Multiple Measures of Creativity

Results from both the feature-counting measure of creativity from prior work and the Likert-scale ratings provided by annotators are largely consistent. While the Likert ratings are subjective, they better capture the creativity in some drawings that combine common or critical features in a novel way, or use a non-visual feature (for e.g. see Fig. 7). Using both together provides a better characterization of creativity.

6 Conclusions and Future Work

This work demonstrates the benefits of early and repeated exposure to examples on creative work. In addition, it suggests that conformity may be the price one pays for these gains, regardless of when examples are seen. Hopefully, these results will encourage designers to seek examples early and often in the design process, when they are most useful.

This experiment also demonstrates a replication (and extension) of creativity studies in an online crowd-sourced environment. Crowd-sourced experiments often offer a lower cost, have a faster time to completion, and provide access to wider populations (Heer and Bostock 2010). This paper's experiment and the labeling tasks took 1 week on Amazon Mechanical Turk. This was possible because the labeling scheme from Marsh et al. (1996) provided this study with a clear taxonomy of features that could be easily labeled by non-experts. We suggest crowd-sourcing as a viable platform both for experiments that do not need participants with specialized skills or background (or modification per participant) and for analysis/labeling tasks that easily verifiable.

This work also raises a number of questions. First, the results of the repeated-exposure experimental condition indicate that the processes of prototyping and learning from examples may be intertwined in a creative task. Further empirical studies could characterize the precise nature of this interaction. Second, this work shows that repeated exposure to examples is beneficial. How does the frequency of (or interval between) such exposures affect this result? Third, designers often spend years acquiring skills and specific domain knowledge. How do such skills and knowledge affect their interaction with examples? Furthermore, similar to cross-cultural effects of prototyping (Kim and Hinds 2012), are effects of examples different in different cultures? Finally, how can the results of this work inform the design of tools that support creative work?

Acknowledgements We thank the Hasso Plattner Design Thinking Research Program for supporting this work.

References

Baayen R, Davidson D, Bates D (2008) Mixed-effects modeling with crossed random effects for subjects and items. J Mem Lang 59(4):390–412

Boroditsky L (2007) Comparison and the development of knowledge. Cognition 102(1):118–128

Buxton B, Buxton W (2007) Sketching user experiences: getting the design right and the right design. Morgan Kaufmann, Amsterdam

Deng J, Dong W, Socher R, Li L, Li K, Fei-Fei L (2009) Imagenet: a large-scale hierarchical image database. In: Computer vision and pattern recognition (CVPR 2009). IEEE conference on, Miami, Florida, USA, pp 248–255

Dow S, Fortuna J, Schwartz D, Altringer B, Schwartz D, Klemmer S (2011) Prototyping dynamics: sharing multiple designs improves exploration, group rapport, and results. In: Proceedings of CHI: ACM conference on human factors in computing systems, pp 2807–2816

Gentner D, Colhoun J (2010) Analogical processes in human thinking and learning. Towards a theory of thinking, Springer Berlin Heidelberg, Berlin, pp 35–48

Heer J, Bostock M (2010) Crowdsourcing graphical perception: using mechanical turk to assess visualization design. In: Proceedings of CHI: ACM conference on human factors in computing systems, pp 203–212

Heit E (1992) Categorization using chains of examples. Cogn Psychol 24(3):341–380

Herring S, Chang C, Krantzler J, Bailey B (2009) Getting inspired!: understanding how and why examples are used in creative design practice. In: Proceedings of CHI: ACM conference on human factors in computing systems, pp 87–96

Jansson D, Smith S (1991) Design fixation. Des Stud 12(1):3–11

Kerne A, Koh E, Smith S, Webb A, Dworaczyk B (2008) Combinformation: mixed-initiative composition of image and text surrogates promotes information discovery. ACM Trans Inform Syst (TOIS) 27(1):5

Kim H, Hinds P (2012) Harmony vs. disruption: the effect of iterative prototyping on teams creative processes and outcomes in the west and the east. In: Proceedings ICIC: international conference on intercultural collaboration. ACM

Lee B, Srivastava S, Kumar R, Brafman R, Klemmer S (2010) Designing with interactive example galleries. In: Proceedings of CHI: ACM conference on human factors in computing systems, pp 2257–2266

Lewis S, Dontcheva M, Gerber E (2011) Affective computational priming and creativity. In: Proceedings of CHI: ACM conference on human factors in computing systems, pp 735–744

Lim Y, Stolterman E, Tenenberg J (2008) The anatomy of prototypes: prototypes as filters, prototypes as manifestations of design ideas. ACM Trans Comput- Hum Interact (TOCHI) 15(2):7

Marsh R, Bower G (1993) Eliciting cryptomnesia: unconscious plagiarism in a puzzle task. J Exp Psychol Learn Mem Cogn 19(3):673

Marsh R, Landau J, Hicks J (1996) How examples may (and may not) constrain creativity. Mem Cognit 24(5):669–680

Newman M, Landay J (2000) Sitemaps, storyboards, and specifications: a sketch of web site design practice. In: Proceedings of DIS: ACM conference on designing interactive systems, pp 263–274

Ritchie D, Kejriwal A, Klemmer S (2011) d. tour: style-based exploration of design example galleries. In: Proceedings of UIST: ACM symposium on user interface software and technology, pp 165–174

Schön D (1985) The design studio: an exploration of its traditions and potentials. RIBA Publications for RIBA Building Industry Trust, London

Smith S, Ward T, Schumacher J (1993) Constraining effects of examples in a creative generation task. Mem Cognit 21(6):837–845

Snow R, O'Connor B, Jurafsky D, Ng A (2008) Cheap and fast – but is it good?: evaluating non-expert annotations for natural language tasks. In: Proceedings of the conference on empirical methods in natural language processing, pp 254–263

Suwa M, Tversky B (1997) What do architects and students perceive in their design sketches? A protocol analysis. Des Stud 18(4):385–403

Ward T (1994) Structured imagination: the role of category structure in exemplar generation. Cogn Psychol 27(1):1–40

Weisberg R (1999) Creativity and knowledge: a challenge to theories. In: Sternberg R (ed) Handbook of creativity. Cambridge University Press, New York, p 226

Yu L, Nickerson J (2011) Cooks or cobblers?: crowd creativity through combination. In: Proceedings of CHI: ACM conference on human factors in computing systems, pp 1393–1402

Part II
Design Thinkers Must Preserve Ambiguity

Impact and Sustainability of Creative Capacity Building: The Cognitive, Behavioral, and Neural Correlates of Increasing Creative Capacity

Grace Hawthorne, Eve Marie Quintin, Manish Saggar, Nick Bott, Eliza Keinitz, Ning Liu, Yin Hsuan Chien, Daniel Hong, Adam Royalty, and Allan L. Reiss

Abstract The impact and sustainability of creative capacity building over time is examined using both neural and psychological approaches. Our research proposes a unique experimental design to test whether creativity can be acquired or learned by an individual over time and how this relates to cognition, behavior, and the brain. In this chapter, we review the background work that focuses on specific cognitive, behavioral, and neural processes that may contribute to creative capacity building. We summarize key components of our experimental design, overview its implementation, and preview early outcomes of intervention research as it relates to the creative capacity building.

G. Hawthorne (✉) • A. Royalty
Hasso Plattner Institute of Design (d.school), Building 550, 416 Escondido Mall, Stanford, CA 94305-3086, USA
e-mail: grace@dschool.stanford.edu; aroyalty@stanford.edu

E.M. Quintin • M. Saggar • N. Bott • E. Keinitz • N. Liu • A.L. Reiss
Center for Interdisciplinary Brain Sciences Research, Stanford University School of Medicine, 401 Quarry Road, Stanford, CA 94305-5795, USA
e-mail: quintin@stanford.edu; saggar@stanford.edu; nbott@stanford.edu; ekienitz@stanford.edu; ningl@stanford.edu; reiss@stanford.edu

Y.H. Chien
Department of Pediatrics, Taipei City Hospital Zhong-Xing, Institute of Biomedical Electronics and Bioinformatics, National Taiwan University, No. 1, Sec. 4, Roosevelt Road, Taipei, 10617, Taiwan
e-mail: dtpedr81@gmail.com

D. Hong
Institute of Biomedical Engineering, National Taiwan University, No. 1, Sec. 4, Roosevelt Road, Taipei, 10617, Taiwan
e-mail: ddh0410@gmail.com

H. Plattner et al. (eds.), *Design Thinking Research*, Understanding Innovation,
DOI 10.1007/978-3-319-01303-9_5,

1 Introduction

In the last 5 years there's been a seismic shift in what prized, must-have skill is needed to navigate a world of increasing complexity. IBM's Institute for Business Values asked 1,500 chief executives in 2010 what leadership competency they valued above all others to face these challenges (Kern 2010). Creativity topped their charts. In 2011, LinkedIn reported that "creative" was the most popular characteristic people used to describe themselves for professional profiles in the United States (Linked in 2011).

The enormous opportunity of creativity is also the underlying challenge because studying the creative process is not easy. As the cornerstone of innovation, creativity is an elusive human characteristic that can make one individual a better design thinker than another. Regardless if you believe that you were born with an endless supply of creativity or not, knowing that you can acquire it through practice has positive implications across all industries and areas of innovation.

Although there have been several recent studies focused on where creativity originates in the brain (Abraham et al. 2012; Aziz-Zadeh et al. 2013; Fink et al. 2012; Green et al. 2012; Jung et al. 2010a; Jung et al. 2010b; Takeuchi et al. 2012), none have focused on this construct as an acquired individual skill, in particular, assessing a person's capacity to become more creative over time, measuring an individual brain's acquisition of creativity, and determining if regular 'exercise' or practice is required to maintain it.

The Hasso Plattner Institute of Design at Stanford University (Stanford d.school) offers classes specifically aimed at enhancing creative capacity through design thinking skill building. As is custom in the academic tradition, instructors must come up with ways to evaluate their student's progression. Often, students are asked to produce deliverables that are then judged by the instructors and classmates. Although this method has academic value, its purely qualitative nature does not allow a formal assessment or measurement on whether the student's creative capacity has been enhanced. **Thus, our primary question is: Can creativity be acquired or learned by an individual over time and if so, does being more creative change an individual's patterns of thoughts and behaviors and even brain functioning?**

Setting out to tackle this question fueled other questions that will be addressed in this chapter. First, we defined a contemporary working definition of creativity as it is conceptualized by the Stanford d.school. Then we asked what human characteristics does creativity relate to on a psychological and biological level? Specifically, what are the cognitive, behavioral, and neural correlates of creative capacity? If creativity can be acquired or improved, does its maintenance require continuous conditioning (like sit-ups for your body) to be retained? What *new* brain and behavioral features are acquired/enhanced by an individual exposed to a creative capacity building instructional course? What *pre-existing* brain, cognitive, personality and behavioral features predict response to a creative capacity building instructional course?

In the first part of this chapter, we review literature related to these questions as background for studying creativity. In the second part of this chapter, we propose an experimental design to study these questions.

This study involved a multidisciplinary team with representation from the fields of design, arts, cognitive and computational neuroscience, psychology, and psychiatry. These diverse perspectives shape the methodology and outcome value. We opted for an intervention aimed to increase creativity called Creative Capacity Building Program along with a control intervention (Language Capacity Building Program). Before and after the intervention, we collected structural and functional magnetic resonance imaging (MRI) brain scans and we administered neurocognitive, personality, and creativity assessments. This hybrid of neuro-imaging methods, neurocognitive assessments, psychological questionnaires and qualitative surveying will allow us to answer the questions stated above.

2 Creativity

Different fields define creativity in various ways. For the purpose of this study, our definition of creativity is inspired by the philosophy of the Stanford d.school while insuring translation to the field of cognitive neuroscience. We define creativity as "*a state of being and adaptation of personal skill sets that enables an individual to synthesize novel connections and express meaningful outcomes*". This definition captures the intersection of three different axes. To determine how creative a person, deliverable or process is, these components can be rated along three continuums from – (a) existing to new/novel, (b) linear to synthesizing, and (c) no value/meaning to meaningful. We propose a visual illustration of these continuums with three axes (Fig. 1a). A deliverable or process with high novelty, meaning, and synthesis is considered highly creative and so is the person responsible for this deliverable or process. This person, the deliverable, or the process falls within the upper right and back zone of the three-dimensional space created by these three axes (the zone in orange in Fig. 1b). This definition of creativity focuses attention to the person and their skills as opposed to process and outcomes as more traditionally defined. The intention of this focus is to better align the skill of creativity to indications that go beyond the possession of creativity into the ability to exercise/apply it.

2.1 *Building Creative Capacity*

The Hasso Plattner Institute of Design at Stanford University (the d.school) was founded in the School of Engineering in 2006 to prepare a generation of innovators to tackle the complex challenges facing the world today. Solutions won't come from any single field, but from collaboration between creative thinkers who can see beyond the way the world is, to the way it could be. The d.school brings together students and faculty from radically different backgrounds to develop innovative, human-centered solutions to real-world challenges. Design thinking is best learned

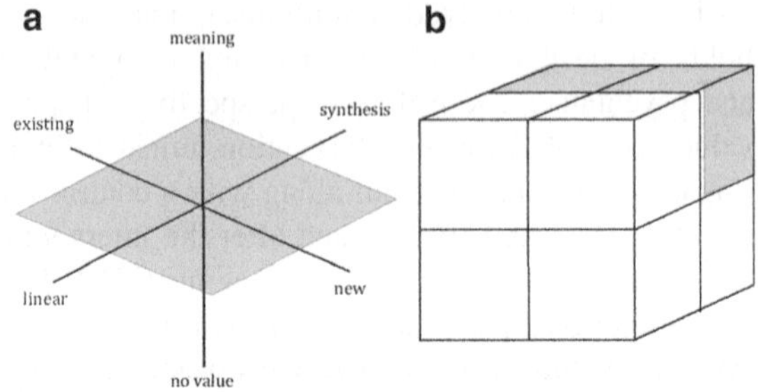

Fig. 1 Visual illustration of our working definition of creativity

by doing, and our classes immerse students in an experiential learning environment. Students cycle rapidly through a series of steps: observe, brainstorm, synthesize, prototype, and implement; repeating as necessary. We focus on the design process because we seek to equip our students with a methodology for producing reliably innovative results in any field. Our focus is on transformative experiences that create innovators rather than any particular innovation.

Creative Gym is a course at the d.school that is devoted to individual skill building. Officially known as ME366, Creative Gym is a graduate level course that has been taught to students and industry executives across many disciplines. The class is an interactive studio where students can build their creative confidence and sharpen their individual design thinking skills through hands-on experiences. Participants engage in unconventional, simple exercises that take them far beyond their everyday experiences in order to train their intuition and push them to think without boundaries in the face of heavy constraints. The Creative Gym experience is an innovative regimen of hands-on exercises and an immersive and interactive experience that is organized around nine core themes that engage our human abilities in intersecting ways.

By introducing participants to fast-paced immersive exercises, the class encourages a potent bias towards action and a deeper understanding of their own personal skills as designers – it gives them a new way to approach life and experience the world that is truly transformative. The resulting bias towards action is their practiced activation of their individual creative capacity.

2.2 *How Does Creativity Relate to Cognition, Behavior, and the Brain?*

Research on creativity and cognitive processes has typically assessed the relationship between performance on neurocognitive tasks and the performance on measures of creativity. Unfortunately, this work gives us little information on the developmental nature of creativity and the factors that contribute to changes in creativity over time. Nonetheless, surveying the literature guides us in choosing which specific aspects of cognition, personality, and the associated neural correlates may contribute to changes in creativity through time.

Our definition of creativity fits with previous research that has described creativity as a human characteristic. Specific aspects of personality and cognition have been associated with creativity. One of the most renowned ways to characterize different personalities is to assert a value to each of five personality traits: the Big Five personality traits (Costa and McCrae 1992). They include openness, conscientiousness, extraversion, agreeableness, and neuroticism. People who obtain higher scores on standardized assessments of creativity, also score high on measures of openness-to-experience (Funrnham 1999; Jung et al. 2010b; Wolfradt and Pretz 2001). Using other classifications of personality, personality traits of efficacy, independence, cognitive control, tolerance and integrity-honesty were associated to being more creative while emotional stability, anxiety, dominance, aggressiveness, and leadership were associated to being less creative (Sanz de Acedo Baquedano and Sanz de Acedo Lizarraga 2012). Given that personality traits, specifically openness, are associated to differences in creativity between individuals, the same traits could also influence changes in creativity over time and are thus important to consider in our intervention study.

Similarly, factors associated to creativity at the cognitive level may influence response to creative capacity building. Cognitive inhibition, which includes the mental ability to focus attention while inhibiting a prepotent response, plays a role in ideational fluency or the ability to generate ideas quickly while intelligence impacts the originality of the ideas that are generated (Benedek et al. 2012). A cognitive strength of creative people is their ability to focus their attention in the most efficient way to respond to task demands indicating flexibility in adjusting focus of attention (Ansburg and Hill 2003). Vartanian et al. (2007) showed that creative potential is associated with an increase attention when no interference is present. For example, creative people respond faster when asked to press a button every time a stimulus appears on a screen and distractions are minimal. However, Vartanian et al. also showed that creative people respond slower on attention task where interference is present such as having to make decisions about a stimulus based on a set of conflicting rules. An oversimplification of these findings would be that creative people have a more flexible way of using their attentional resources. Fluency, flexibility, and originality are included within divergent thinking abilities, which are more related to creativity than general intelligence (Kim 2008). Thus, research has shown that attention, inhibition, fluency, flexibility are specific neurocognitive factors that influence creative capacity building. These factors are included within the broad concept of executive functions. Executive functions are higher-order cognitive processes required to organize thoughts and behavior. They include inhibition, attention, working memory, planning, organization, and verification. They impact general cognitive ability or general intelligence and divergent thinking, which are also important factors to consider as correlates of creativity.

Recently, a series of neuroimaging studies have focused on the neural correlates of creativity in the brain (Abraham et al. 2012; Aziz-Zadeh et al. 2013; Fink et al. 2012; Green et al. 2012; Jung et al. 2010a; Takeuchi et al. 2010). Although limited to a static view of creativity at one fixed point in time, these studies provide a starting point for our research. In one of the first studies linking neuroanatomy and creativity, Jung et al. (2010b) established a link between cortical thickness in different areas of the brain and creativity. They operationalized creativity using pre-established Creativity Achievement Questionnaire (CAQ; Carson et al. 2005) and divergent thinking tasks and found that thickness of posterior cingulate, left orbitofrontal, and right angular gyrus was related with scores on these measures of creativity. In another similar study, Diffusion Tensor Imaging (DTI) was used to examine the relation between creativity (operationalized using divergent thinking tasks) and white matter integrity in the brain (Jung et al. 2010a). The authors reported that fractional anisotropy (FA) values were inversely related to creativity in the left inferior frontal white matter, which plays a role in executive functioning. FA values are measures of axonal coherence and myelination. Thus, an inverse relation between FA values and creativity suggests decreased myelination in participants scoring higher on creativity. Based on these and other anatomical correlates of creativity, we also plan to obtain DTI scans and assess whether DTI metrics including FA are associated with creative capacity building.

Apart from structural predictors of creativity, researchers have also investigated brain connectivity predictors for creativity. For example, a recent study by Takeuchi et al. (2012) examined the brain's functional connectivity while participants are at rest (i.e. not involved in any active task; a.k.a. rFC) to find out which neural circuits of the resting brain are related with creativity (operationalized using divergent thinking tasks). They found that higher scores on creativity tasks were associated with higher connectivity between medial prefrontal cortex (mPFC) and posterior cingulate cortex (PCC) in the brain. The mPFC and PCC regions are usually considered to be key components of default mode network (DMN), which deactivates as soon as a participant actively performs a task requiring significant cognitive processing. The authors suggest that participants achieve higher creativity through increased interaction within DMN, thereby combining ideas represented in different regions in the network. Thus, the DMN is also important to consider in our study of creative capacity building.

In search of neural correlates of creativity, researchers have also examined functional activity in the brain while participants were performing tasks that required creative solutions (Arden et al. 2010; Dietrich and Kanso 2010 for review). Most researchers have mainly focused on divergent thinking aspect of creativity. However, a few have also attempted to find neural correlates of creative problem solving in real world-like situation. For example, Fink et al. (2012) tested whether the exposure to other people's ideas would impact the creativity of participants engaged in a divergent thinking task. The authors simulated brainstorming sessions and found that cognitive stimulation by exposing common or moderately creative ideas improved the participants' creativity. Temporo-parietal brain regions (in right hemisphere) were found to be sensitive to such cognitive stimulation, by possibly playing the role of integrating new information and previous knowledge.

Altogether, neuroimaging research suggests that many brain regions are associated with creativity, many of which are in frontal brain regions. This has been demonstrated by examining brain structure, the DMN, anatomical connectivity within the brain (via DTI techniques), or brain activation associated with performance on divergent thinking tasks. Our study will include these techniques to identify neural correlates of creative capacity building.

3 Experimental Design

Using a scientific method to approach the question of creative capacity building, we propose a unique and novel experimental design (see Fig. 2), where we collected data related to creativity, cognition, behavior, and the brain before (time 1 or T1) and after (time 2 or T2) a creative capacity building program based on the Creative Gym class offered at the d.school. We also included a control group receiving a Chinese language and character drawing learning intervention (or Language Capacity Building Program). Both interventions lasted 5 weeks with weekly meetings of 2 h per week. Participants were randomly assigned to either intervention. Thus, there were two groups of participants. The groups were matched on age, gender, and IQ. Following the intervention and time 2 data collection, the groups crossed-over to receive the other intervention, i.e. participants in the creative capacity building group were assigned to the Chinese language and character drawing learning intervention and vice versa. The second set of interventions was followed by assessments and brain scans at time 3 (or T3). This experimental design allows us to compare changes in cognition, behavior, and brain function from time 1 to time 2 between groups (creative capacity building program vs. control) and assess whether these changes are maintained over time (from time 2 to time 3). Thirty-six people completed our first testing point (T1 in Fig. 2). Participants were randomly divided into two groups of 18. Figure 2 shows how many participants completed each portion of the experimental design.

Our recruitment surpassed what we had initially proposed in the grant. To recruit participants, we sent out advertisements for our study to family and friends and listservs for groups on Stanford University's campus and in the area. Ninety-two people replied with an interest for the study. We then contacted each of these 92 people to verify if they met inclusion criteria for the study. Thirty-six of the 92 people met the criteria for inclusion in the study. Participants were not eligible if they were fluent in Chinese and knew how to write Chinese characters. Exclusion criteria also included left-handedness, non-removable metallic devices, a history of neurological or serious psychiatric disorder, pregnancy, and some types of medications. Participants were aged between 18 to 39 years old and were available for the entire study, which included two 5-weeks trainings sessions and MRI brain scans, neurocognitive, and behavioral assessments at three time points. In total, this represents a 30–35 hour time commitment for each participant over approximately 15 weeks. In total, 6 (of the initial 36) participants did not complete the entire research protocol for personal reasons. Each participant terminating their participation was debriefed by one of the experimenters.

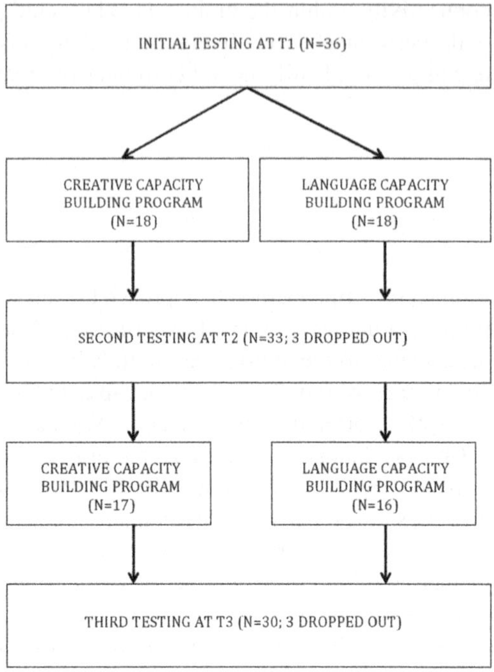

Fig. 2 Visual overview of our experiment design

4 Interventions

All participants underwent both training sessions during the study, what differed between groups was the order they received the interventions.

4.1 Creative Capacity Building Program

For this study, we created a short version of the d.school "Creative Gym" class. The adaptation of the class engaged participants in best of breed, fast-paced immersive exercises from each of the original course's nine themes. In an open design studio setting, participants worked individually on unconventional, fast-paced immersive exercises. The hands-on activities used everyday office supplies as materials. Participants were able to reflect on their exercises through post-activity reflections and viewing other participants' work. Each training session was comprised of a diverse variety of activities that were made known to the participants as they were assigned to them.

4.2 Language Capacity Building Program

In this intervention, participants learned how to pronounce some basic Chinese words and the general rule of writing Chinese characters. This was done in a group setting to recreate the shared environment atmosphere of the design studio. Participants copied and practiced Chinese characters with regular pens and by using traditional Chinese brush and ink (calligraphy) on tracing books. This part of the intervention gave an opportunity for hands on exercise while minimizing creative generation of ideas. To motivate participants, the instructor briefly introduced Chinese history and Chinese calligraphic history.

5 Assessments and Data Collection

To provide a glimpse of the richness of multiple dimensions of the collected data, we provide a brief introduction to the most important measures that were used at various time points (Fig. 3).

5.1 Assessing Creativity

As an index of creativity, we used the Figural Torrance test of creativity thinking (fTTCT), the Design Thinking Creativity Test (DTCT), the Creative Agency and Confidence Questionnaire (CACQ), and the Creative Achievement Questionnaire (CAQ). The fTTCT, CTCT, and CACQ were administered at each time point (T1, T2, T3) while the CAQ was only administered at the first time point. The fTTCT includes three picture-based exercises to assess five mental characteristics: fluency, resistance to premature closure, elaboration, abstractness of titles, and originality. It also provides scoring for the following creative strengths: emotional expressiveness, internal visualization, storytelling articulateness, extending or breaking boundaries, movement or action, humor, expressiveness of titles, richness of imagery, synthesis of incomplete figures, colorfulness of imagery, synthesis of lines or circles, fantasy, and unusual visualization (Torrance 1974). The DTCT, developed at d.school by Hawthorne et al., is a companion assessment to the TTCT that reflects problem solving needs in the twenty-first century by testing an individual's ability to exercise their creativity effectively in real world scenarios. The CACQ, also developed at d.school by Royalty et al., is a self-reported scale to evaluate subjects' creative self-efficacy and perception of their own creative capacity. Finally, the CAQ is a reliable and valid measure of creative productivity across ten domains including visual arts, music, creative writing, dance, drama, architecture, humor, scientific discovery, invention, and culinary arts (Carson et al. 2005).

AREA ASSESSED	NAME OF MEASURE	DESCRIPTION	TIME POINT T1	T2	T3
General intelligence	Wechsler Abbreviated Scale of Intelligence (WASI)	This test measures verbal knowledge, reasoning ability, concept formation, visuospatial problem solving, and abstract reasoning	X		
Executive function	Delis-Kaplan Executive Function System (D-KEFS)	This test is used to measure a variety of verbal and nonverbal executive functions including attention, inhibition, and flexibility	X	X	X
Creativity	Figural Torrance Test of Creativity Thinking (TTCT)	Uses picture-based exercises to assess: fluency, resistance to premature closure, elaboration, abstractness of titles, and originality	X	X	X
Creativity	Design Thinking Creativity Test (DTCT)	Uses a scenario-based activity to assess application of creative characteristics	X	X	X
Creativity	Creative Confidence Questionnaire	Self-assessment to capture creative self-efficacy and agency	X	X	X
Creativity	Creativity Achievement Questionnaire (CAQ)	Measures creative productivity across ten domains including visual arts, music, creative writing, dance, drama, architecture, humor, scientific discovery, invention, culinary arts	X		
Personality	NEO-FFI-3	NEO-Five Factor Inventory (NEO-FFI) was used to assess the personality types for each participant	X		
Neuroimaging	Structural Brain Imaging (MRI)	Provides information about the anatomy of the brain: volume, thickness, and surface area of different brain structures	X	X	X
Neuroimaging	Functional Brain Imaging (fMRI)	Provides dynamic information about the functioning of the brain while the participant is performing a task	X	X	X
Neuroimaging	Resting State Brain Imaging (fcMRI)	Method for evaluating regional interactions that occur when a participant is not performing any explicit task (a.k.a. Default-Mode-Network)	X	X	X
Neuroimaging	Diffusion Tensor Imaging (DTI)	Used to infer connectivity amongst brain regions	X	X	X

Fig. 3 Summary table of assessments

5.2 Assessing Cognition and Behavior

Given previous evidence of personality traits, specifically openness, on creativity, we used the NEO-Five Factor Inventory (NEO-FFI-3, Costa and McCrae 2010) to assess the personality traits for each participant. It has 60 questions (12 per domain) evaluating the following personality traits: extraversion, agreeableness, conscientiousness, neuroticism, and openness to experience. This measure was only administered at the first time point because personality traits are thought not to vary through time.

We administered an IQ test to verify that groups did not differ on this factor to ensure that potential group differences in creative capacity would not be attributable to an IQ difference between the groups. IQ is also stable through time and was thus only assessed at the first time point. We used the Wechsler Abbreviated Scale of Intelligence (WASI-II, Wechsler 2011). This test includes two subtests of verbal knowledge, reasoning ability, and concept formation and two subtests of visuospatial problem solving and abstract reasoning.

Other neurocognitive factors that may influence creative capacity building include attention, inhibition, fluency, and flexibility. These factors can be included within the broad concept of executive functions. We hypothesized that some components of executive functioning may change as creative capacity changes and we thus assessed these functions at each time point (T1, T2, T3). We used three subtests of the Dellis-Kaplan Executive Functions Systems Test (DKEFS, Delis et al. 2001). The verbal fluency and design fluency subtests measure fluency, flexibility, executive memory, and inhibition. The color-word interference test measures inhibition and flexibility.

5.3 *Assessing the Neural Correlates of Creativity*

In order to find the neural correlates of creativity, we employed a series of neuroimaging techniques and collected data at all the three time points. Starting with the anatomical correlates of creativity, we acquired structural MRI data that will allow us to measure changes (if any) in brain structure associated with creativity training. We also plan to use the anatomical data (at T1) to predict improvements in creativity after training (at T3). Second, using High Angular Resolution Diffusion Imaging (HARDI) technique we also acquired anatomical connectivity information. Using anatomical correlates we plan to not only identify the regions associated with creative capacity, but also to use these correlates in conjunction with other assessments for predicting and understanding subject-to-subject variability in the dataset. For example, we can find out whether certain brain areas or networks are correlated with increased creative capacity and how they change over time.

To find the functional correlates of creativity, we used functional brain imaging (fMRI). This method provides dynamic information about the functioning of the brain while the participant is performing a task in the scanner. In our case we used the following two tasks to investigate neural correlates of increased creative capacity. Both of these tasks were implemented using a novel MRI-safe drawing tablet that was designed especially for our project by Dr. Bob Dougherty (Director, Center for Cognitive and Neurobiological Imaging, Stanford University).

The first task, the trail-making test, is an fMRI adaptation of a neurocognitive assessment commonly used outside the scanner to measure visual attention and flexibility or shifting, which have been associated with creative capacity. The task consists of two parts in which the participant is instructed to connect a set of 12 dots as fast as possible while still maintaining accuracy. It can provide information about visual search speed, scanning and speed of processing, mental flexibility, and executive functioning. In part A, participants connected numbers in sequential order (1, 2, 3, etc.). In part B, participant connected an alternating sequence of numbers and letters (1, A, 2, B, etc.). Participants used the MR-safe drawing tablet to complete this task.

The second fMRI task is a novel task that was designed specifically to measure the participant's creative capacity. Saggar et al. created an experimental task inspired by the Pictionary™ game (Designers: Angel and Everson 1985) and adapted to imaging constraints. In condition A, participants were shown a word (usually a verb) and were asked to draw it using the MR-safe drawing tablet so that the word/action can be recognized. This is an open-ended task where participants can draw as much or as little as they deem appropriate. This task involved several components of the creative capacity building intervention implicitly, for example, it involved building, synthesis, navigation, etc. In condition B of the task, participants were asked to draw a control word, "zig-zag".

Altogether, using a myriad of behavioral and neuroimaging tests we planned to explore the correlates of creativity capacity building in an exhaustive manner.

6 Implications

Preliminary findings have been generated, that we invite the reader to interpret with caution until the findings are published in a peer-reviewed scientific article. The results comparing group performance on neurocognitive and behavioral tasks suggest that the creativity training intervention did successfully improve some aspects of creativity and had an impact on some components of executive functioning (Bott et al. 2013; Kienitz et al. 2013). Changes in creativity and executive functioning do not seem to be dependent on personality types. Preliminary findings from our imaging tasks also indicate that the creativity training intervention was associated with changes in the neural resources associated with executive functioning and imagination.

7 Looking Ahead

New findings on the brain basis and sustainability of creative capacity and design thinking skills will provide completely unique information to the field. This information has the potential to profoundly influence our understanding of human brain and behavior links underlying design thinking during the phenomenon of innovation.

By understanding if a person can improve their creative capacity and whether it is retained, metrics and method of increasing creative and instructional effectiveness can be improved. By diving deeper into the question of individual skill building and creativity retention over time, we can discover the true value of design thinking during the phenomenon of innovation and know better how to teach it.

References

Abraham A, Pieritz K, Thybusch K, Rutter B, Kröger S, Schweckendiek J et al (2012) Creativity and the brain: uncovering the neural signature of conceptual expansion. Neuropsychologia 50(8):1906–1917

Angel R, Everson G (1985) Pictionary. Angel Games, Inc for fist edition. Current Publisher, Hasbro

Ansburg PI, Hill K (2003) Creative and analytic thinkers differ in their use of attentional resources. Pers Individ Differ 34:1141–1152

Arden R, Chavez RS, Grazioplene R, Jung RE (2010) Neuroimaging creativity: a psychometric view. Behav Brain Res 214(2):143–156

Aziz-Zadeh L, Liew S-L, Dandekar F (2013) Exploring the neural correlates of visual creativity. Soc Cogn Affect Neurosci 8(4):475–480

Benedek M, Franz F, Heene M, Neubauer AC (2012) Differential effects of cognitive inhibition and intelligence on creativity. Pers Individ Dif 53(4):480–485

Bott NT, Kienitz E, Quintin EM, Saggar M, Hawthorne G, Royalty A, Reiss AL (April 2013) Creativity training enhances self-directed attention and information processing. Poster presentation at the Cognitive Neuroscience Society Annual Meeting. San Francisco

Carson SH, Peterson JB, Higgins DM (2005) Reliability, validity, and factor structure of the creative achievement questionnaire. Creat Res J 17(1):37–50

Costa PT, McCrae RR (1992) Manual of the revised NEO personality inventory. Psychological Assessment Resources, Odessa

Costa PT, McCrae RR (2010) NEO five-factor inventory-3. Psychological Assessment Resources, Lutz

de Acedo S, Baquedano MT, de Acedo S, Lizarraga ML (2012) A correlational and predictive study of creativity and personality of college students. Span J Psychol 15(3):1081–1088

Delis DC, Kaplan E, Kramer JH (2001) Delis–Kaplan executive function system. Pearson Education, San Antonio

Dietrich A, Kanso R (2010) A review of EEG, ERP, and neuroimaging studies of creativity and insight. Psychol Bull 136(5):822–848

Fink A, Koschutnig K, Benedek M, Reishofer G, Ischebeck A, Weiss EM, Ebner F (2012) Stimulating creativity via the exposure to other people's ideas. Hum Brain Mapp 33 (11):2603–2610

Funrnham A (1999) Personality and creativity. Percept Mot Skills 88(2):407–408

Green AE, Kraemer DJM, Fugelsang JA, Gray JR, Dunbar KN (2012) Neural correlates of creativity in analogical reasoning. J Exp Psychol Learn Mem Cogn 38(2):264–272

Jung RE, Grazioplene R, Caprihan A, Chavez RS, Haier RJ (2010a) White matter integrity, creativity, and psychopathology: disentangling constructs with diffusion tensor imaging. PLoS One 5(3):e9818

Jung RE, Segall JM, Jeremy Bockholt H, Flores RA, Smith SM, Chavez RS, Haier RJ (2010b) Neuroanatomy of creativity. Hum Brain Mapp 31(3):398–409

Kern B (2010) What chief executives really want. *Bloomberg businessweek.* Bloomberg L.P., New York

Kienitz E, Bott NT, Quintin EM, Saggar M, Hawthorne G, Royalty A, Reiss AL (April 2013) Creativity training enhances creative persistence and increased abstract connections. Poster presentation at the Cognitive Neuroscience Society Annual Meeting. San Francisco

Kim KH (2008) Meta-analyses of the relationship of creative achievement to both IQ and divergent thinking test scores. J Creative Behav 42(2):106–130

Linked in (2011) http://blog.linkedin.com/2011/12/13/buzzwords-redux/?trk=corpblog_1212

Takeuchi H, Taki Y, Sassa Y, Hashizume H, Sekiguchi A, Fukushima A, Kawashima R (2010) White matter structures associated with creativity: evidence from diffusion tensor imaging. Neuroimage 51(1):11–18

Takeuchi H, Taki Y, Hashizume H, Sassa Y, Nagase T, Nouchi R, Kawashima R (2012) The association between resting functional connectivity and creativity. Cerebral Cortex 22 (12):2921–2929

Torrance EP (1974) The torrance tests of creative thinking-norms technical manual research edition-verbal tests, forms A and B- figural tests, forms A and B. Personnel Press, Princeton

Vartanian O, Martindale C, Kwiatkowski J (2007) Creative potential, attention, and speed of information processing. Pers Individ Dif 43:1470–1480

Wechsler D (2011) The Wechsler abbreviated scale of intelligence, 2nd edn. Pearson Education, San Antonio

Wolfradt U, Pretz JE (2001) Individual differences in creativity: personality, story writing, and hobbies. Eur J Pers 15(4):297–310

Acting with Creative Confidence: Developing a Creative Agency Assessment Tool

Adam Royalty, Lindsay Noelle Oishi, and Bernard Roth

Abstract Universities around the world are quickly introducing new learning models aimed at developing creativity and innovation in students. A leading model is the experiential teaching of design thinking as a creative problem solving process aimed at enhancing students' creative confidence. Although these programs exist, little is known about student outcomes. Furthermore, the criteria by which we evaluate student "success" is not well defined because these programs almost uniformly have ambiguously stated learning objectives. This research uses qualitative and quantitative data to capture and categorizes successful outcomes by examining alumni of these programs. Based on these data is a scale that measures creative agency, a fundamental outcome of teaching design thinking.

1 Introduction

This research focuses on a new model of teaching creativity and innovation through a design process, often called design thinking (Cross 2007). In particular, this paper focuses on one of the leading institutions in this field, The Hasso Plattner Institute of Design at Stanford University ("d.school"). Stanford University and the d.school have garnered a reputation for producing successful entrepreneurs and innovators according to a recent Stanford Entrepreneurship Survey, and there is widespread interest in replicating its educational methodology. For example, universities in Japan, Chile, Malaysia and China have founded or are in the process of creating educational programs modeled on the d.school. Corporations are also interested in teaching design and creativity; in Europe, Germany's design group Palomar 5 and PICNIC in The Netherlands have created ambitious programs inspired by the d.school's educational methods. While the outcomes of these endeavors have not

A. Royalty (✉) • L.N. Oishi • B. Roth
Stanford University, Stanford, USA
e-mail: adam@dschool.stanford.edu; loishi@stanford.edu; broth@stanford.edu

H. Plattner et al. (eds.), *Design Thinking Research*, Understanding Innovation,
DOI 10.1007/978-3-319-01303-9_6, © Springer International Publishing Switzerland 2014

yet been fully evaluated, previous educational trends have demonstrated that expensive, culturally grounded practices rarely work as expected when imported to new international contexts.

One reason that imported educational methods tend to "fail" in new contexts is that the expected results are often imperfectly understood and specified (e.g., "math achievement" or "innovation"). To help those interested in the d.school and design thinking avoid this pitfall, we first sought to describe the actual outcomes of the d. school's educational practices using qualitative and quantitative research methods rather than anecdotal reports or case studies alone. Such impact questions are often motivated by goals of program evaluation, that is, identifying what is pedagogically effective for the sake of promoting or propagating it, and what is ineffective for the purpose of improving it. At this level, only *what* works is important; *why* it works is unimportant. On a more theoretical level, however, we were also interested in the psychological mechanisms underlying whatever learning happens within an educational setting and curriculum such as the d.school. Our research, therefore, occupies the intersection of psychological research and program evaluation: we wished to identify the psychological constructs relevant to effective learning experiences within a particular program (the d.school), with the goal of designing evaluations that would reflect a common learning framework that would apply across educational settings.

Naturally, describing the d.school's impact is an impractically large research question. Any reasonably complex educational institution has myriad impacts and innumerable contributing variables. Furthermore, as institutions like the d.school are non-traditional and interdisciplinary, they often do not have specific or easily measurable target outcomes for students. This is quite different from a graduate program in Physics that teaches students to do research in one of several subfields, each with a known set of standards and criteria with which to determine alumni achievement. At the d.school, the only concrete outcome identified in the mission statement is the production of "future innovators" (d.school 2004). Therefore, a first step in investigating the impact of the d.school is selecting targeted outcomes for which research will be appropriate and useful. One outcome of the d.school, for example, could be graduates' romantic relationships; however, from an educational perspective this is of relatively minor interest. Given that the prevailing motivations for studying and imitating the d.school stem from its apparent contributions to business and the economy (see, for example, the *Design Ladder* developed by the Danish Design Centre), *we chose to focus on students' professional outcomes after graduation.*

A second step in designing research for understanding the impact of the d.school is narrowing the field of variables we examine as characteristic of the institution. The space of potential variables is vast, including, for example, geographical location, student population, instructor characteristics, material resources, and individual course curricula. Because the d.school is inherently concerned with learning and teaching (Beckman and Joyce 2012), and because its unique pedagogy is its most salient feature, we chose to study *student outcomes related to the d. school's signature problem-solving approach, design thinking.*

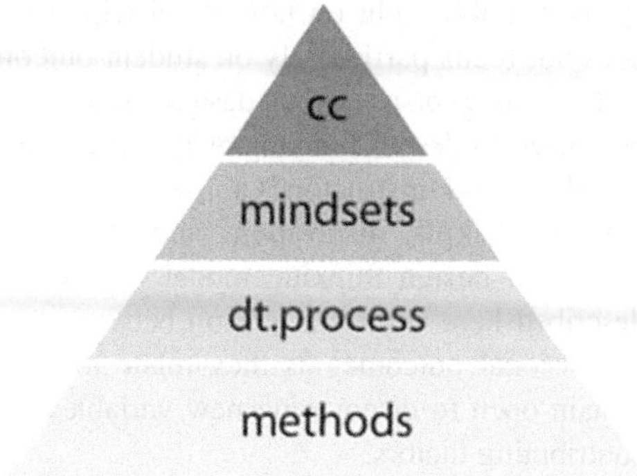

Fig. 1 Model of creative confidence (Rauth et al. 2010)

2 Background

While design thinking has been defined in different ways (Buchanan 1992), (Cross 2007), d.school courses are based on a common pedagogy that focuses on an overall *process* with five core *constructs*: Empathy, Define, Ideate, Prototype and Test. On a more specific instructional level, nearly all d.school courses involve techniques such as interviewing, brainstorming, and rapid prototyping. Design thinking as it is taught at the d.school goes beyond explicit pedagogy. The goal of the d.school is to develop future innovators, who, in addition to performing discrete observable skills, also have a characteristic set of attitudes and dispositions that propel them toward creative activity and achievement.

According to a framework of a d.school teaching model (Fig. 1), mindsets and an overall sense of creative confidence are built on top of repeated practice and success with discrete techniques (Rauth et al. 2010) such as the design thinking process (dt.process) and its various associated methods. These mindsets include a bias towards action, radical collaboration, and being human centered. Other dispositions, not specified in the model but commonly promoted at the d.school, include constant reframing and rapid iteration.

The literature about design thinking outcomes is fairly sparse. A great amount of design methodology research focuses on engineering or design teams and how they perform (Eris 2004), (Brereton et al. 1996). Because the d.school does not train designers, but rather helps students from all disciplines work in a more creative way, the existing design research is not a sufficient base.

Kolb's learning model (Beckman and Barry 2007) connects how the design process helps students develop a sense of integrated thinking. This is very valuable, but we are interested in learning how deeper behavioral aspects are affected, and a more holistic view of how individual skills and techniques learned work together to create the overall problem-solving-approach of design thinking. There is also research exploring teaching models that are similar to design thinking, such as Leifer's (1996) work evaluating Product-Based-Learning Education, Gerber's examination of Design-Based Learning (Gerber 2011), and the Cambridge-MIT Institute (Lucas and Cooper 2004; Lucas and Cooper 2005). Though such work sheds valuable light on how the d.school teaching mechanism affects students, it does not focus particularly on student outcomes beyond the learning experience.

In deciding on a research design, we also kept in mind that the outcomes believed to be related to design thinking as it is taught at the d.school are based on the d.school founders' and instructors' a priori intuitions and beliefs. It is quite likely that graduates' skills, dispositions and attitudes are not fully specified by either the traditional design thinking model or that shown in Fig. 1. Therefore, the studies described here exhibit a tension between our a priori research focus on graduates' professional outcomes as they relate to the d.school pedagogy, and our desire to remain open to discovering new variables of interest, in both outcomes/impact and contributing factors.

We chose individual alumni as the unit of analysis versus organizations that have d.school graduates because the d.school's mission is that of personal behavior change and developing individual attitudes, not redefining how companies operate. That said, knowing how individuals who apply d.school learnings within a corporate context does inform the impact of design thinking on a company (Gerber and Carroll 2012).

As confidence to think and act creatively came up frequently in the qualitative data, an initial psychological construct we used to characterize this phenomenon was self-efficacy (Bandura 1977). Self-efficacy refers to an individual's belief in his or her abilities within a particular domain, in this case, creative problem-solving. Another design thinking institution has done some work on how self-efficacy may be related to design pedagogy (Jobst et al. 2012). Self-efficacy, however, is only one part of a larger construct, agency, which Bandura (1982) defined as the means by which "people can effect change in themselves and their situations through their own efforts," (p. 1175). Agency includes not only self-efficacy, but also beliefs about the world, context, physical and emotional states, social support, and other factors. Though it is more complex, we felt that agency better reflected the multifarious nature of the creative competencies that many d.school graduates exhibited. We defined creative agency as individuals' capacity to effect change in themselves and their situations to support successful creative problem-solving.

Given that there is little research on the outcomes of programs that involve teaching design thinking through problem-based learning methodologies, we decided to take a wide-to-narrow approach starting with exploratory qualitative studies (an open-ended survey with follow-up interviews) and an initial conceptual model based on our observations. To test the model's key psychological construct, agency, we did a pilot test using a new scale, revised it based on the data, and then tested a second version of the scale across multiple populations (study 3).

3 Study 1: Alumni Survey

Although we reported the background, methodology and initial results for Study 1 in last year's report (Royalty et al. 2012), for this year, we analyzed additional data and applied a new coding scheme to several questions. The preliminary goal of this continued work on last year's data was to generate a foundational set of descriptive results regarding d.school alumni, so that stakeholders, prospective students and the larger design education community can better understand the alumni population. The second, more research-oriented purpose of these additional analyses was to address the following *new* research questions:

1. To what extent does the d.school affect alumni professional outcomes (i.e. career choice)?
2. To what extent are alumni professional activities related to d.school pedagogy?

As this study was exploratory in nature, there were no a priori hypotheses. For the sake of completeness, we re-report the methods in the following section, but interested readers can go straight to the results.

3.1 *Method*

3.1.1 Participants

We defined a d.school alumnus as any person who had taken at least one d.school course, where a course was defined as a quarter (ten weeks) of unit-bearing instruction. Based on student records, the participants we surveyed were a close approximation of the entire population (approximately 670 alumni). The response rate was 28 % for a final sample of 184. Among those reporting gender, 56 % were men (n = 73). Nearly all participants had been graduate students at Stanford University (some had finished their degrees and others had not) and were affiliated with various schools and programs, including Business, Law, Arts & Sciences, Education, Medicine and Engineering. As the d.school officially offered courses beginning in 2006, participants had been out of school and employed full-time from zero to a maximum of 5 years. The majority (83 %) had graduated in 2008 or after.

3.1.2 Procedure

Given the large number of potential participants in this study, for ease of data collection and analysis, we chose a survey methodology and used online survey software (Qualtrics). Alumni were identified through student records and contacted via personal email as well as a general Facebook announcement (it is possible that some people received more than one notification or invitation to participate in the study). Invitations were sent in May 2011 and the survey was open for 2 weeks. Participants were free to take the survey at a time and place of their convenience. There was no explicit incentive to participate; however, respondents had the opportunity to join a mailing list that would notify them of upcoming alumni events.

Table 1 Selected survey questions on outcomes of d.school pedagogy

Item	Response type
What are the most important effects or outcomes from your experience from the d. school?	Short paragraph
In the last month, how often did you apply what you learned through the d.school in your professional life?	5-point Likert scale: not in the last month, once or twice, once/week, 2–3 times/week, almost every day
What are some examples of how you applied what you learned through the d.school in your professional life in the last month?	Short paragraph

3.1.3 Materials

All of the questions on the survey were designed specifically for this research. Participants were asked about the current and previous occupations and occupational plans, how they applied what they learned at the d.school in their professional lives (Table 1), and demographic questions. So as not to bias respondents towards reporting specific techniques or general dispositions, we did not use the term "design thinking" in any questions; instead, we referred to "what you learned at the d.school". The survey questions also did not offer any definition or examples of "applying" what was learned at the d.school; respondents were free to respond based on their own interpretation of this term.

3.2 Results

3.2.1 Alumni Professional Outcomes and Their d.school Experiences

By comparing participants' self-reported planned occupation when they began their graduate studies with their current occupation, we found that among those who had a planned occupation, a majority experienced a career change. Table 2 shows the frequency breakdown of all respondents, including those who had career plans and those who did not when they began graduate study. Looking at only those participants who did have a planned employment field and/or role when they began at Stanford ($n = 122$), 62.3 % of these ($n = 76$) reported a difference between the job/career they had planned on when they began their graduate study, and their current employment.

A subset of participants ($n = 70$) reported a career change and also answered the question, "Thinking about the difference (if any) between your expected occupation when you started at Stanford, and your current occupation, how much did your experience at the d.school play a role in this change?" The mean response was 2.47 ($SD = 1.46$), which corresponds to approximately halfway between "A moderate amount" (scale point 2) and "A lot" (scale point 3).

Table 2 Categories and frequencies of career plans and outcomes for d.school alumni

Category	Frequency	Percent of total (%)	Mention d. school	Percent of category mentioning d.school (%)
Career change	76	41.50	26	34.20
No career change	46	25.10	6	13.00
No original plan	59	32.20	3	5.10
No current job given	2	1.10	0	–
Total	183	100.00	35	19.10

Of the 76 career changers, 34.2 % (n = 26) explicitly (and spontaneously) mentioned their experiences at the d.school as a reason for this outcome. Even among those who did not report a career change or did not have an original plan, some spontaneously mentioned their d.school experience as having an impact on their professional path.

To understand any potential role that the d.school had on alumni career choice and professional outcomes, responses were coded from the question, "If your current job is different from what you expected to do when you started at Stanford, what factors played a role in this change?" Among participants who reported working in an occupation that they had not planned on when they matriculated, the most common reason was a change in professional interests or desires (Table 3). Furthermore, approximately half of those who reported a change in interests or desires for their job/employment (19 out of 37) explicitly mentioned the d.school as a reason for this change.

For those who mentioned the d.school as a reason for their change in occupation, the overall theme was that various aspects of the d.school experience (e.g., a particular class, teamwork) enhanced their self-understanding or their understanding of professional roles. Table 4 gives some examples of these open-ended responses.

To understand how specific pedagogical elements of the d.school's instructional methods may have affected alumni professional outcomes, we focused on the open-ended questions; "What are the most important effects or outcomes of your experience at the d.school?" and "What are some examples of how you applied what you learned at the d.school in your professional life in the last month?". As not all participants responded to these questions, the number of respondents for this analysis was 131.

From these responses, we coded for elements of the d.school design process, defined a priori. We also included teamwork as an a priori coding category, given its centrality to the design thinking learning process. We also coded for dispositions that correspond to key tenets of the d.school curriculum. "Creative confidence" is a new term that we used to describe a theme of feeling comfortable with creative endeavors and a sense of ability and self-efficacy in the creative domain (Table 5).

Table 3 Reasons for occupational change among d.school alumni

Reason for occupational change	Frequency (%)	Number mentioning d.school (%)
Change in interests/desires	37 (20.2 %)	19 (51.4 %)
No response/vague response	21 (11.5 %)	3 (14.3 %)
Could not get desired job	9 (4.9 %)	1 (11.1 %)
Unexpected opportunity	4 (2.2 %)	2 (50 %)
Wanted to make impact	3 (1.6 %)	1 (33.3 %)
Money/necessity	2 (1.1 %)	0
Total (all career changers)	76 (100 %)	26 (34.2 %)

Table 4 Example responses to question about change in occupation relationship of alumni professional activities to d.school pedagogy

Theme	
Self-understanding	"It just feels like I can bring more of myself to work and after I had a taste of that experience at the d.school, it was hard to go back"
	"I was taught a method of problem solving that fit better who I am (a creative person) than the traditional engineering way of solving problems"
Specific d.school experience	"I went in looking to play a business role (i.e., strategy) on creative teams. . . I ended up playing more of a hybrid role. Time on teams at the d.School gave me an opportunity to play exactly the type of role that I wanted"
	"I took Needfinding, a class in Product Design, and found my passion in applying empathy to understand problems and finding new opportunities"
Understanding of professional opportunities	"I was introduced to the possibility of designing products that help impoverished people by the d.school"

3.3 Discussion

Re-analyzing the survey data with a particular focus on career changes and how alumni viewed the contribution of d.school pedagogy revealed moderate evidence for the claim that educational methods at the d.school have an impact on graduates' professional outcomes. Across all participants, nearly twenty percent brought up the d.school as a factor in their professional paths and current occupation. It is important to note that these participants mentioned the d.school's influence without prompting – the question about why they were working in their current occupations did not mention the d.school, and was placed before (and on a separate page from) the question that specifically asked about the role of the d.school. Furthermore, the results show that alumni perceive their d.school experience as a causally contributing factor in their occupational choices.

As participants varied in what they found most significant about their experience at the d.school, we cannot conclude that it was, in fact, d.school pedagogy rather than other factors that drove these outcomes. The data on alumni use of

Table 5 Frequency of d.school alumni mentioning d.school pedagogies

Component of d.school pedagogy	Frequency (n) mentioning	Percentage of all respondents (n = 131) (%)
Empathy	56	43
Define	32	24
Ideate	48	37
Prototype/test	49	38
Teamwork	24	19
Creative confidence	33	25
Comfort with risk, ambiguity, change, or failure	32	24
Bias towards action	14	11

fundamental design thinking methods, however, do support a hypothesis that alumni apply d.school lessons in professional settings. Of the six a priori categories that we coded for, empathy, ideation and prototyping were the most commonly mentioned. This is unsurprising as these techniques are in sharp contrast with more traditional work methods, and apply in many situations. The methodology of "defining" a problem was mentioned less frequently, probably because it is a more subtle, procedural step that can be difficult to describe. It was, however, somewhat surprising that only a fifth of participants talked about teamwork, as this is central to nearly all d.school experiences.

Arguably the most important outcome of the survey was the emergence of several new coding categories. We view these as overall psycho-behavioral "dispositions"; in particular, the themes of creative confidence, comfort with seemingly negative states (such as failure or ambiguity), and having a bias towards action (rather than intellectual reflection or rationalization) were frequently observed. In order to better understand how alumni may have developed these holistic dispositions, in addition to more explicit design thinking/problem-solving skills, we decided to do in-depth follow-up interviews with a subset of the alumni surveyed in Study 1.

4 Study 2: Alumni Interviews

As Study 2 was described in the previous report (2011), we will just briefly review the methodology and major findings in order to describe the larger research arc. Sixteen d.school alumni (ten women) from Study 1 were interviewed in person or via teleconference (Skype). Seven participants currently worked in business (including healthcare, technology and entertainment), three were self-employed or entrepreneurs, three worked in education, two in consulting or research, and one was in engineering. The interviews were semi-structured and were 45–100 min.

The interviews probed how "successful" alumni viewed what they had learned at the d.school and how that may have influenced both their career choices and their professional activities. By "successful", we merely meant those who said that they chose to, and were able to, use what they learned at the d.school (which they defined for themselves) in their working life. On a high level, we wanted to see how former d.school students described their learning path and their subsequent behaviors in order to better understand how the d.school may be contributing the formation of "future innovators". More specifically, we focused on their observable behaviors and dispositions, and how alumni perceived the d.school as shaping these outcomes.

The overall result of the interviews was a better understanding of how some d. school alumni develop and demonstrate creative confidence. Beyond simply learning basic design thinking processes and techniques, participants who reported using what they learned at the d.school had a strong desire and confidence to actually apply skills and methods in their workplace. This can be a daunting and risky endeavor in some traditional work contexts, where normal or known behaviors are expected and job responsibilities are tightly codified. In the strongest case – seen in several interviewees – creative confidence in the professional realm even extended to using the design thinking process to launch or change one's entire career. For example, one participant left her job in engineering to pursue teaching, and explicitly stated that it was what she learned at the d.school that gave her the confidence to "try out" or prototype a new career, even though it meant taking significant professional risks.

Another, unexpected example of how alumni demonstrated creative confidence that came up in multiple interviews was how they built creative environments around themselves at work and in their personal lives. This often took the form of manipulating physical space to allow for more creative working styles, and in some cases, making the space similar to the interior of the d.school. Some participants also reported teaching colleagues about design thinking as a way to enable better collaborations and facilitate their own creativity. Across most of the alumni we interviewed, the rich descriptions of how they used what they learned at the d.school went far beyond simple applications of basic techniques or tools such as brainstorming or rapid prototyping. Instead, their stories reflected an overall approach to work and life that was informed by dispositions to solve problems in an innovative manner, and propelled by the confidence to do so in multiple ways.

Successful alumni seemed to share a set of behavioral patterns and dispositions that went beyond what we had captured in Study 1. Although we could observe those characteristics in interviews, we wanted to define a single psychological construct that could be measured accurately and efficiently. In Study 3, we attempted to narrow and quantify what was observed in the survey and interviews, by creating a scale that attempted to measure the new construct of creative agency.

5 Study 3: Competency-Based Creative Agency Scale (CBCA)

In developing a scale measuring the impact of a design-thinking based pedagogy, we revisited the data from the surveys and interviews and coded for key competencies frequently demonstrated by successful alumni. Again, we defined "successful" as those who reported remembering and using "what they learned at the d.school" (self-defined) the most within our sample. We then created the scale directly from those factors. The questions were worded in an abstract manner rather than by referring to specific scenarios, in order to limit the influence of previous experience or gender differences. The key factors that emerged were:

Sources (gathering information from external sources)
Comfort (with ambiguity)
Mastery (of one's own creative process)
Environment (developing creative environments)
Anti-perfectionism (reducing a sense that everything must be perfect)
Prototyping (developing a culture of prototyping)
Perseverance (increased in the face of failure)
Facilitation (confidence to lead a creative process)
Openness (to changes in thought, direction, beliefs, et cetera)
Process (being able to describe one's own creative process)
Creative Output (solving problems in creative ways)

5.1 *Method*

5.1.1 Participants and Procedure

1. *Current d.school students (General).* The scale was implemented in the Autumn quarter of 2011 with current d.school students drawn from two classes. The first was a large, introduction to design methodology course (n = 72) (95 % of class) and the second was a more specialized course exploring the intersection of design and society (n = 13) (76 % of class). The questionnaire was given at the beginning (September) and end (December) of the academic quarter and no incentives were offered for participation. Of the 85 pretest participants, 68 completed the posttest.
2. *Current d.school students (Education).* Thirteen students in a small, Education-focused d.school course took the questionnaire as a pretest and posttest at the beginning and end of the Spring 2012 academic quarter. The pretest was given at the beginning of (April) and the end (June) of the Spring quarter; eight students took the posttest. A $15 gift card was offered for participating.

Table 6 Items on the competency-based creative agency scale

Item	Construct
Find sources of creative inspiration not obviously related to a given problem	Creative idea sourcing
Effectively work on a problem that does not have an obvious solution	Comfort with ambiguity
Change the definition of a problem you are working on	Openness
Shape or change your external environment to help you be more creative	Building creative environments
Share your work with others before it is finished	Anti-perfectionism
Try an approach to a problem that may not be the final or best solution	Prototyping
Continue work on a problem after experiencing a significant failure	Perseverance after failure
Help others be more creative	Creativity facilitation in others
Identify and implement ways to enhance your own creativity	Mastery of creative process
Explicitly define or describe your creative process	Knowledge of creative process
Solve problems in ways that others would consider creative	Successful creative problem-solving

3. *Non-d.school students (Education)*. Twelve Education students who were not taking the Education-related or any other d.school course prior to or during Spring 2012 were recruited as a control group for the d.school/Education group. These students completed the questionnaire online.
4. *Workshop participants: Business executives*. Participants in a 3-day design thinking workshop for business executives received an email with a link to an online questionnaire, 2 weeks prior to the workshop in March 2012. 57 participants took the pretest (98 % response rate). To date, six participants have taken the posttest, which was also provided via email, 10 weeks after the workshop. Those who took the post-test received a t-shirt.
5. *Workshop participants: Business executives and teachers*. Participants in a 3-day design thinking workshop for in-service teachers and business executives took the questionnaire on paper, at the beginning and end of the workshop ($n = 62$) (52 % response rate at pretest), which took place in May, 2012.

5.1.2 Materials

There were 11 to 14 Likert-scale questions on the CBCA scale; all were created for this study. Eleven items concerned self-assessed confidence related to a set of general competencies in the area of creative problem-solving (Table 6). The CBCA-related Likert-scale items were presented as a group and preceded by the question, "How confident are you that you could…". Unlike the items in Study 3, the new questions were written to be sufficiently general so as to apply to nearly any situation or professional domain. There were five response categories: "Not at

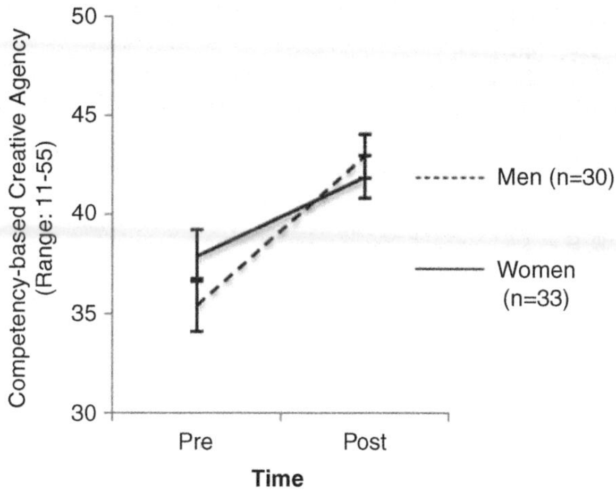

Fig. 2 Mean competency-based creative agency, by time and gender

all confident" (scale point 1), "A little confident" (2), "Moderately confident" (3), "Very confident" (4), and "Completely confident" (5). There were no "Don't Know" or "Not Applicable" response choices; however, participants were free to leave any item blank.

After implementing the 11-item questionnaire with the first group of participants, all of whom were d.school students, we added three new items on topics relatively unrelated to competency-based creative agency. These items concerned comfort with technology, artistic ability, and goal achievement. They were added to provide data on whether any apparent outcome of d.school study was specific to the design thinking competencies we found in earlier studies, or a more general positive impact.

5.2 *Results*

While there was no overall effect of gender, there was some evidence of a trend for gender to interact with time, such that women started out with higher CBCA but ended at a similar mean score, $F(1, 53) = 3.04, p = .087$ (Fig. 2). The effect size for this possible interaction was, however, quite small, with a partial eta-squared of .05.

There were too few participants in the non-d.school student "control" group for statistical analysis; however, among the eight who took both pretest and posttest, there was no apparent change in mean CBCA. At pretest, the control group mean was 35.25 ($SD = 5.31$); at posttest, the mean was 36.12 ($SD = 3.68$). Figure 3 plots the means and standard errors for the d.school students versus non-school students at pretest and posttest.

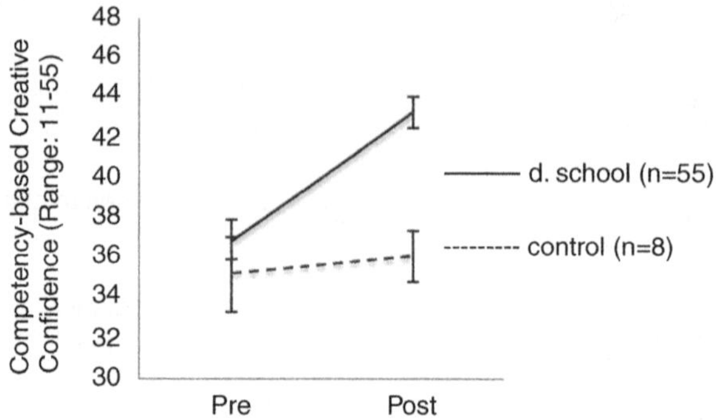

Fig. 3 Mean competency-based creative agency, by time and group

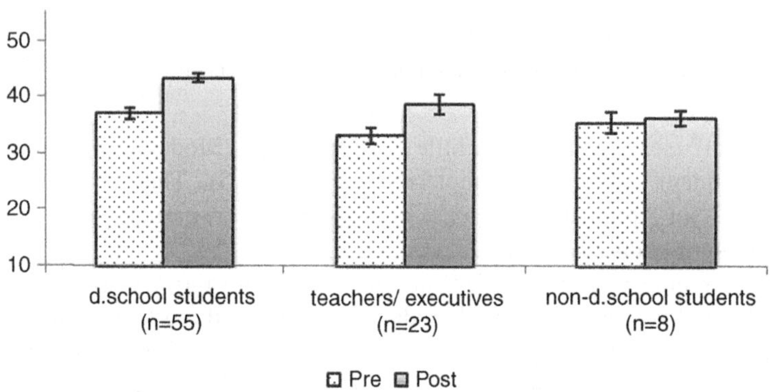

Fig. 4 Mean competency-based creative agency, by sample and time

For the 23 teacher/executive workshop participants for whom we have both pretest and posttest scores, mean CBCA at pretest was 32.91 ($SD = 7.13$) and mean CBCA at posttest was 38.65 ($SD = 8.50$). A paired-samples t-test showed a significant effect of time, $t(22) = -4.50$, $p < .001$. Figure 4 shows the pretest and posttest scores for each participant – only three decreased in their CBCA scores, while the remaining 20 increased.

5.3 *Discussion*

Study 3 tested a new scale, competency-based creative agency, with three main samples: d.school students, similar students not taking d.school classes, and teachers and executives who took a 3-day design thinking workshop. As the scale items were based on empirical research on the applied skills of successful creative problem-solvers who were d.school alumni, it was hypothesized that d.school experience would be associated with greater CBCA (see Fig. 4). The results provided evidence for this relationship, as d.school students increased in CBCA from the beginning to the end of the course, across two different academic quarters and in three different d.school classes. In one of those quarters, a small group of non-d.school students also took the scale; they did not show any significant change in CBCA. While a larger control group would, of course, provide better support for the hypothesis, these data do suggest that the increase in CBCA seen among d. school students was not merely due to maturation or demand characteristics of the scale items.

Results from the teacher/executive sample also supported the hypothesis that d. school experience increases CBCA. Although the design thinking workshop was only 3 days, participants reported a significant increase in CBCA from the beginning to the end. Data from this sample also showed that the scale could be used with a non-students sample and retain its internal validity as well as its usefulness in showing change over time.

It is interesting to note that the average pre-test score of the teachers and executives was slightly lower than the average pre-test score of the non-d.school students, which in turn was marginally lower than the average pre-test score for d. school students. There was, however, an apparent difference between the mean starting CBCA for d.school students and teachers/executives, with d.school students reporting higher CBCA at pretest. Taking a design thinking course shows willingness to take risks by studying a topic that is relatively new and not a regular part of any degree program; a higher pretest score among these students, then, is not surprising.

6 Implications

The three studies reported here have led us toward a larger model of how individual creativity develops and is subsequently expressed. One of the original concerns of our research was to investigate how the d.school and other educational institutions can foster the kinds of skills that lead to real-world innovations. Our initial model of creative outcomes, as observed in our research, includes four constructs: self-efficacy, agency, output and impact (Fig. 5).

Creative self-efficacy is belief in one's own creative abilities, and it is developed in many ways, such as mastery experiences (successfully completing creative activities) and vicarious experiences (learning from others who are creative). Self-efficacy is the most important part of creative agency, which goes beyond mere belief in oneself and extends to real-life application of creative problem-solving. Those who have high creative agency are capable of producing creative output – real work, artifacts, and ideas. Creative output determines creative impact, this is the change made due to a creative effort.

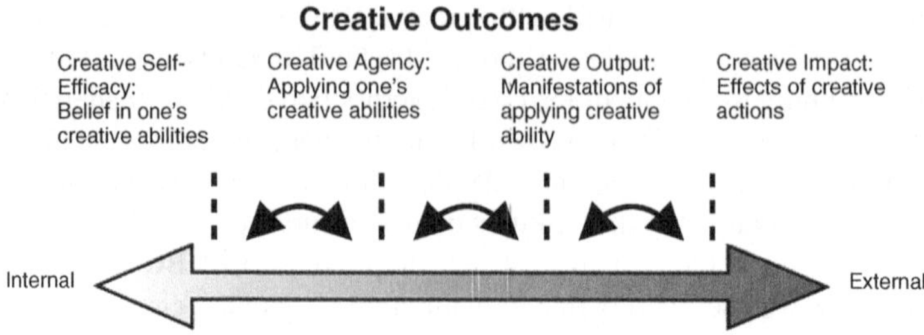

Fig. 5 A model of multiple creative outcomes

This model is not intended as a linear or temporal path, as there is multi-path feedback among the nodes. Self-efficacy naturally leads to greater agency, which may result in greater output, which strengthens both self-efficacy and agency. This can result in more creative output, which can strengthen agency that can lead to even more creative self-efficacy. The model is also not intended to be exhaustive: given the work being done to better understand structures of the creative brain, we may soon include segments to the left of creative self-efficacy, representing more internal structures. Likewise, there may be multiple segments to the right of creative impact, when one considers the impacts of not just individuals, but also institutions and other large systems.

7 Work in Progress

We are currently in the early pilot stages of three studies aimed at the creative output node of our model, that is, the ideas, work, artifacts, procedures and other "deliverables" that creative agents make.

8 Conclusions and Future Work

Understanding the impact of any educational institution or method is a daunting task. By focusing on the unique problem-solving approach taught at the d.school, design thinking, and grounding our work in an understanding of the characteristics that d.school alumni report, we have begun to provide evidence for the impact of d. school pedagogy on a newly identified construct, creative agency. Instead of focusing on more traditional outcome measures, such as salaries, patents, or awards, we chose to look at the competencies, capacities and habits that those demonstrated by alumni who say that they use what they learned at the d.school in their work and lives.

What emerged from our qualitative research is a sense that many d.school alumni are equipped with a strong sense of their own ability to be creative in problem-solving contexts (self-efficacy), and that this propels them to act to change their own thinking and their environments in order to perform better as creative problem-solvers (creative agency). These actions are characterized by multiple competencies, first learned and practiced at the d.school, and then carried into the workplace. These include beliefs (anti-perfectionism, failure as opportunity), knowledge (of one's creative process), dispositions (towards prototyping, radical collaboration), and skills (teaching others, shaping one's own environment). We also observed that these capacities were largely domain-agnostic, and were reported by teachers, engineers, entrepreneurs, and even chefs in equal measures.

In our quantitative research, we moved towards an efficient and accurate way to represent and measure the outcome we observed among the alumni, by designing and testing two scales. While the first did not function well as a singular-construct scale, we did see interesting results, such as, more experienced d.school students said that they were more comfortable starting a company than students with less d. school experience. Our second scale, which built upon the lessons of the first, was more successful at providing a valid and reliable measure of the impact of d.school pedagogy. By testing it with several populations, we found initial support for its use beyond the d.school. In the near future, we hope to repeat our studies with larger and more robust control groups, and implement the scale in other educational programs and institutions, both similar to and different from the d.school.

With more evidence for the validity and reliability of the Competency-based Creative Agency scale, we would like to eventually expand the research to its potential predictive validity, its neurological correlates, and ways to use it in controlled experiments. For example, using the CBCA in the workplace may yield insights into how workers with high creative confidence are recognized for their creativity and the benefit they provide to their companies (predictive validity). We are also collaborating with another research team on fMRI studies that incorporate the CBCA scale. Finally, to truly test the effects of the d.school pedagogy, we would need to run experimental studies that manipulate what we believe are key aspects of design thinking as it is learned at the d.school, and then measure how these components affect CBCA. This will take us closer to understanding the contribution of the d.school to educational methods, as well as how the successes of the d.school may be not only repeated, but also enhanced, at other organizations throughout the world.

References

Bandura A (1977) Self-efficacy: toward a unifying theory of behavioral change. Psychol Rev 84(2):191–215

Bandura A (1982) Self-efficacy mechanism in human agency. Am Psychol 37(2):122–147

Beckman S, Barry M (2007) Innovation as a learning process: embedding design thinking. Calif Manage Rev 50(1):25–56

Beckman S, Joyce CK (2012) Reflections on teaching design thinking to MBA students. In: Business as an agent of world benefit conference, 2–5 June 2009

Brereton MF, Cannon DM, Mabogunje A, Leifer L (1996) Characteristics of collaboration in engineering design teams: mediating design progress through social interaction. In: Dorst K, Christiaans H, Cross N (eds) Analyzing design activity. Wiley, Chichester

Buchanan R (1992) Wicked problems in design thinking. Des Issues 8(2):5–21

Cross N (2007) Designerly ways of Knowing. Birkhauser Verlag AG, Boston

d.school (2004) dschool.stanford.edu. Retrieved 10 July 2013 from https://dschool.stanford.edu/

Eris O (2004) Effective inquiry for innovative engineering design. Kluwer, Boston

Gerber E (2011) Extracurricular design-based learning: preparing students for careers in innovation. Int J Eng Edu 28(2):317–324, 2012

Gerber E, Carroll M (2012) The psychological experience of prototyping. To appear in design studies (forthcoming)

Jobst B, Köppen E, Lindberg T, Rhinow H, Moritz J, Meinel C (2012) The faith-factor in design thinking: creative confidence through education? In: Design thinking research studying co-creation in practice. Springer, Heidelberg, pp 35–46

Leifer L (1996) Evaluating product-based learning education. White Paper. Center for Design Research, Stanford University, Palo Alto. Retrieved 29 Feb 2008

Lucas WA, Cooper SY (2004) Enhancing self-efficacy to enable entrepreneurship: the case of CMI's connections. MIT Sloan school of management working paper, 4489–04

Lucas WA, Cooper SY (2005) Measuring entrepreneurial self-efficacy. In: EDGE conference: bridging the gap: entrepreneurship in theory and practice, Singapore, 11–13 July 2005

Rauth I, Köppen E, Jobst B, Meinel C (2010) An educational model towards creative confidence. 1st Proc. ICDC, Kobe

Royalty A, Oishi L, Roth B (2012) 'I Use it Everyday': pathways to adaptive innovation after graduate study in design thinking. In: Design thinking research measuring performance in context. Springer, Heidelberg, pp 95–105

How Design Thinking Tools Help To Solve Wicked Problems

Julia von Thienen, Christoph Meinel, and Claudia Nicolai

Design thinking is a method of understanding problems and producing innovative, compelling solutions. The problems design thinkers tackle have come to be called "wicked" in the literature. But how does design thinking help to solve wicked problems? This article takes a psychological perspective to analyze the particular challenges wicked problems bring about. Many of the design thinking tools will then turn out to provide effective support for meeting exactly these challenges.

If design thinking is a means to solve problems – what problems is it good for? Obviously, it is not made to help physicists compute precise mathematical solutions. Neither does it help the industry to make their standard products a little faster, smaller or shinier than before.

The problems design thinkers take on are typically very much down to earth. Much has been written about the characteristics of design thinking problems. In general, the notion is that design thinking tackles *wicked problems* (Buchanan 1992; Lindberg et al. 2009, 2012).

Rittel and Webber (1973) have provided a first and highly influential discussion of wicked problems. They provided ten key notions, starting with the following.

"1. There is no definite formulation of a wicked problem" (p. 161).

J. von Thienen (✉)
Hasso-Plattner-Institut an der Universität Potsdam, Prof.-Dr.-Helmert-Str. 2-3, 14482 Potsdam, Germany
e-mail: Julia.vonThienen@hpi.uni-potsdam.de

C. Meinel
Internet Technologies and -Systems, Hasso Plattner Institute for Software Systems Engineering, Campus Griebnitzsee, Postdam 14482, Germany
e-mail: christoph.meinel@hpi.uni-potsdam.de

C. Nicolai
Hasso Plattner Institute at the University of Potsdam, School of Design Thinking, Prof.-Dr.-Helmert-Str. 2-3, D-14482 Potsdam, Germany
e-mail: claudia.nicolai@hpi.uni-potsdam.de

H. Plattner et al. (eds.), *Design Thinking Research*, Understanding Innovation,
DOI 10.1007/978-3-319-01303-9_7,

Consider mathematical problems. They may be stated once and forever. Or think of industrial requests. These may be quite rigid, e.g., how to make a car five hp stronger. Wicked problems, however, are so multi-faceted that many different problem views are viable. Imagine, for instance, a case from psychology: There is an unhappy family. The husband is mostly off at work. The wife is overstrained by her job, the housework and restless kids. The children lack joy and warmth at home. What is the problem? Many answers are conceivable.

"2. Wicked problems have no stopping rule" (p. 162).

When facing a mathematical problem, there may be a point when *the* solution is found – or when it becomes clear that there is no solution at all. An industrial request for making a car 5 hp stronger is fulfilled when the car is 5 hp stronger. But at what point should an unhappy family stop trying to change things for the better? It is often a matter of gut feeling when to give up hope completely or when to make peace with the situation reached.

"3. Solutions to wicked problems are not true-or-false, but good-or-bad" (p. 162).

A mathematical solution is right or wrong. But how about the unhappy family? Has the problem been "solved correctly" once the husband has given up his job, became a stay-at-home dad and now tries to unburden his wife? What if the kids find gratification in hobbies away from home and rarely expect anything of their parents any more? What if the couple splits up so that their ceaseless quarrels end? Evidently, there are many ways to counter the unhappiness-problem. And there may well be approaches which are not exactly false, but rather bad solutions.

Thus far, it may be compelling to say that design thinkers tackle wicked problems as described by Rittel and Webber (1973). Yet, the ten criteria they formulated were spelled out for the field of politics. And design thinking problems are not like political problems in all respects. Consider, for instance, the following criteria formulated by Rittel and Webber.

"5. Every solution to a wicked problem is a 'one-shot operation'; because there is no opportunity to learn by trial-and-error, every attempt counts significantly" (p. 163).

This may be said about a government which decides over war or peace. But why should the unhappy family not try out something new? How about a role change for a week? How about new formats of work-sharing? How about involving the grandparents?

"10. The planner has no right to be wrong" (p. 166).

Again, this makes sense when a regime imposes decisions on community members who have no choice but to live with the consequences. Design thinking, however, typically brings about solutions which people may adopt or refuse. Users will adopt design thinking solutions only if convinced by them and reject them otherwise.

In the process of working out a solution, design thinking allows for many trials and many errors. Solutions need to be shaped carefully so that users will find them thoroughly satisfying. Core mottos such as ***fail early and often***, ***bias for action*** or ***embrace experimentation*** reflect this emphasis on trial and error and learning.

Therefore, not all ten criteria provided by Rittel and Webber (1973) seem to be equally pertinent to the problems found in design thinking. We shall limit ourselves to discussing three characteristics, still obviously assimilating Rittel and Webber's ideas.

With the condensed three criteria in mind, let's ask a simple question: What do the problems common in design thinking actually demand of the people or teams who tackle them? And how can design thinking be of help in that regard?

1 There are Many Stakeholders Involved; and Their Problem Views may be Quite Volatile

Consider the unhappy family. If you ask each family member what the problem is you may hear quite diverse answers. The children may find their parents "less cool" than their friend's parents. The husband may complain that his efforts to earn money for the family are not valued enough by his wife. The wife may feel she is left alone with all responsibilities and duties; she needs more support. And the kids are a little too loud and too messy.

Interestingly, these problem views need not be stable. Because there is no correct or false framing of the problem, people may change their views as they encounter new perspectives. For example, one friend of the wife might convince her that it is all her husband's fault. Isn't he egoistical to head for self-fulfilment at work while she has no choice but to struggle with laundry baskets and cleaning rags? Another friend might convince her that the problem lies with her and not with the husband. Why does she feel responsible for everything? She should involve the kids more frequently and take up her old hobbies.

So, a design team working on problems of this kind needs to:

Create a stable problem view!

This is very much in line with one of the famous d.mindsets. "Craft clarity[!] Produce a coherent vision out of messy problems. Frame it in a way to inspire others and to fuel ideation" (Hasso Plattner Institute of Design at Stanford 2011).

Given that problem views of stakeholders need not be coherent and may actually be quite volatile, it is an important job to reach a stable interpretation of the problem. Otherwise, every solution you prepare may be considered promising by the stakeholders in one moment and may be discarded in the very next. As problem views change, people also change their understanding of what a solution should achieve. Thus, they introduce new criteria for evaluating your proposals.

Chances for establishing a stable problem view obviously increase if you:

Integrate differing perspectives!

If your problem view ignores the perspective of a central stakeholder, he may rebel against it. Imagine designing something delightful for the kids of the unhappy family and then having the parents consider it an impertinence. The end result: they will simply ban it.

Generate a new and inspiring insight!

Why is the problem still around? Apparently, the common views of the problem have not forwarded lasting solutions. Provide a new interpretation of the problem that warrants new hope, introduce a new angle!

Last but not least:

Communicate your idea in a catchy way!

If people don't understand your take on the problem, they cannot adopt it.

Design thinking has much to offer in reaching stable problem views. Some of the common tools are these.

Creative reframing helps to keep in mind that many different problem views are possible. And it's crucial to find a good one.

Multidisciplinarity makes different outlooks on a problem likely even when all team members have observed the same scene. Since differing outlooks may seem fruitful, evidently there is no single "correct" interpretation.

Interviewing for empathy and ***assuming a beginner's mindset*** help to explore the perspectives of different stakeholders and bring up new angles.

Mottos such as ***go for quantity*** or ***encourage wild ideas*** propel beyond the problem views and solutions that are out there already, which have failed.

Feedback in different project phases and from differing audiences helps to reach understandable and compelling problem statements (as well as solutions).

Prototyping eases communication with users and allows you to test how sweeping your approach is.

Story-telling is a means to communicate problem views compellingly.

2 Solutions are not Correct or False - They Just Meet the User's Needs More or Less Well

Once again, consider the unhappy family. Imagine if a therapist would appear and act as the ultimate connoisseur of human nature. According to his analysis, the husband just needs to exert more authority at home. Under his instruction the kids should help their mother more often so that she will be unburdened and therefore able to appreciate her husband again.

The family may try, but what if happiness and love fail to materialize? Maybe the husband doesn't enjoy ordering others around. Maybe the wife is quite an emancipated woman who sets value on equal roles. Maybe the kids long for parents who pay more attention to their concerns. Not even a professional like a therapist can claim to possess an ultimate connection to "the truth". Nothing more and nothing less can be done than exploring people's needs and inventing strategies to meet them.

Therefore, a central challenge is to

Ask for needs, not for the truth!

Again, design thinking introduces many tools to facilitate the job.

Both the ***focus on human values*** and the ***empathize mode*** immediately draw attention to needs.

And, of course, there are explicit empathy techniques. For example, ***empathy maps***, ***bodystorming*** or ***analogous empathy*** allow you to consider more than what people express in words in order to understand their fundamental concerns.

Also, when ***multiple design teams*** work in parallel on the same challenge they typically come up with variegated and often times compelling solutions. So you learn that, apparently, there is no single "true" solution.

3 The Process of Problem Solving Needs to be Productive without Ultimate Criteria of Success

Imagine you want to help the unhappy family. Congratulations if you don't impose your "solution" on them right after shaking hands for the first time. A solution to their problem will emerge only after exploring carefully what world they live in and what they need there (***defer judement***).

But if you don't have a solution ready in your mind how can you and the family be optimistic that a good idea will pop up in a couple of days, weeks or months? After all, the family may have tried for years to help themselves. . . without success.

Surely, individual tools may be of value which support a central step in the problem solving process.

Merge needs and insights!

Remember, all you can do regarding wicked problems is to embrace needs. And a fresh outlook seems essential since the old problem views were obviously fruitless. So, you also need new insights. Design thinking has powerful tools to ease the job.

For example, a ***point of view madlib*** or ***how-might-we questions*** push towards generative problem statements and thereby propel towards gratifying solutions.

But there ought to be a path leading towards needs and insights. And there ought to be a path leading from ideas to ready and tested solutions. The process of problem solving may involve many steps, so it would be helpful to have at one's disposal a reliable strategy or a plan.

Yet, it is also clear what this strategy needs to avoid. Obviously it would be a bad idea to start formalizing the problem in order to deduce "the optimal solution". We are not dealing with mathematics here. There are no right or wrong or objectively optimal solutions. And every formalization would be arbitrary.

Instead, the process needs to be flexible enough to basically handle all kinds of topics, leaving room to explore and articulate needs no one may have thought of before.

Thus, there are a couple of challenges to be met. As a design thinker you will need means or heuristics to:

Structure without formalizing!
Focus without formalizing!
Explore widely and stop wisely!

Design thinking offers loads of means to channel the process of problem solving in a productive direction without forestalling any concrete decisions.

The ***design thinking process model*** provides orientation without dictating concrete steps.

A ***game-plan*** allows teams to structure their approach – but not too rigidly.

Time constraints and a ***gong*** provide pragmatical stopping criteria that push the process forwards.

Maxims like ***encourage wild ideas*** or ***saturate white boards,*** tools such as ***watching out for extreme users***, applying ***powers of ten*** or ***brainstorming*** and setup features such as ***multidisciplinary design teams*** help to explore widely.

Maxims such as ***bias for action*** or ***visualize*** and tools like ***prototyping*** or ***voting after brainstorming*** help to focus without formalizing.

Iteration helps to circle in on promising ideas.

All in all, design thinking offers ample help to solve wicked problems of a liberal type where you may fail and experiment first to become all the more successful later on. It does so by establishing mindsets and offering tools which save you from the impossible task of finding "the correct problem view" or "the optimal solution". Instead, attention is drawn to needs which await their fulfillment. New interpretations of the problem are advanced which take into account the perspectives of different stakeholders and which help to look at the matter from a new angle, since the old problem views turned out to be blind alleys. Finally, a lot of tools are provided to drive the process of problem solving in a productive direction – making sure the process remains flexible, spirited and unrestrained by arbitrary formalizations.

References

Buchanan R (1992) Wicked problems in design thinking. Des Issues 8(2):5–21

d.school, Hasso Plattner Institute of Design at Stanford (2011) Bootcamp Bootleg. http://dschool.stanford.edu/wp-content/uploads/2011/03/BootcampBootleg2010v2SLIM.pdf. Accessed on 29 Nov 2012

Lindberg T, Noweski C, Meinel C (2009) Design thinking. Zur Entwicklung eines explorativen Forschungsansatzes zu einem überprofessionellen Modell. Neuwerk, Zeitschrift für Designwissenschaft 1:47–54

Lindberg T, Köppen E, Rauth I, Meinel C (2012) On the perception, adoption and implementation of design thinking in the IT industry. In: Plattner H, Meinel C, Leifer L (eds) Design thinking research. Studying co-creation in practice. Springer, Berlin, pp 229–240

Rittel HWJ, Webber MM (1973) Dilemmas in a general theory of planning. Policy Sci 4:155–169

Part III
All Design Is Re-design

How Prototyping Helps to Solve Wicked Problems

Birgit Jobst and Christoph Meinel

Abstract This article discusses prototyping as a design-thinking tool to solve wicked problems. It will do so by exploring the relation between three factors considered important in the process: (1) prototyping skills, (2) wicked problem solving competence and (3) wicked problem self-efficacy. We shall propose a model regarding their interaction and suggest approaches for testing it as means to enhance wicked problem solving competence by improving prototyping skills.

1 Introduction

The core of design is problem solving. Most of the world's really important problems which must be solved are wicked ones. Because wicked problems involve several stakeholders, with different interests this makes it difficult to solve them (Rittel and Webber 1973). That is the reason why wicked problems, as for example child poverty, lack of resources, and urban delinquency, are often old problems, which have not been solved until now. There is a need to develop human-centered and holistic solutions to soften the consequences of a globalised world with increased complexity of a radicalized modernity for further generations (Giddens 1996) where the number of wicked problems will grow. One approach to develop the needed innovation is the education in innovative thinking and human-centered development of ideas and the solving of wicked problems (wicked problem solving competence). As one possible way to solve wicked problems some authors mention design thinking as a wicked problem solving technique (Buchanan 1992; Lindberg

B. Jobst (✉)
Hasso Plattner Institute, Potsdam, Germany
e-mail: Birgit.Jobst@hpi.uni-potsdam.de

C. Meinel
Internet Technologies and -Systems, Hasso Plattner Institute for Software Systems Engineering, Campus Griebnitzsee, Postdam 14482, Germany
e-mail: christoph.meinel@hpi.uni-potsdam.de

H. Plattner et al. (eds.), *Design Thinking Research*, Understanding Innovation,
DOI 10.1007/978-3-319-01303-9_8,

et al. 2010) Buchanan argues that design is fundamentally concerned with the particular and there is no science of the particular.

We suppose that wicked problem solving competence could be enhanced via good prototyping skills (Wiese and Wiese 2012) and by wicked problem self-efficacy, both are important conditions for the development of innovative ideas.

What could effect a possible improvement of wicked problem solving competence within the setting of the schools of design thinking? This and related questions are matters that current research has not yet widely examined. We suppose that one way to foster wicked problem solving competence is via an enhancement of prototyping skills and wicked problem self-efficacy, which is addressed by schools of design thinking. In this contribution, we discuss how factors as wicked problem self-efficacy, wicked problem solving competence and prototyping skills interact with each other.

2 The Model

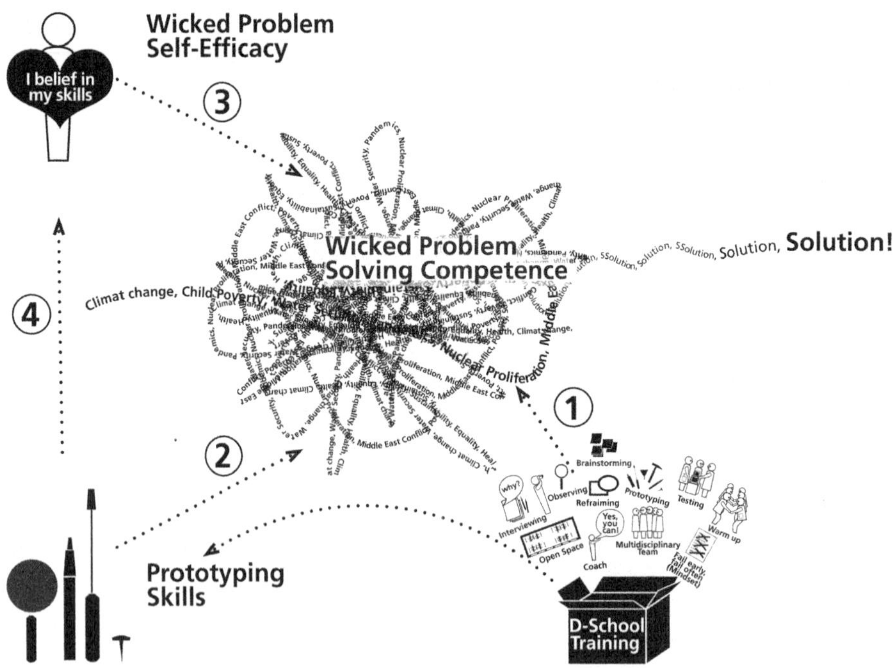

In the following, we will present the three factors of the model -wicked problem self-efficacy, prototyping skills and wicked problem solving competence- and how these factors interact.

3 Design Thinking Provides a Toolbox for Solving Wicked Problems (Arrow 1)

3.1 Wicked Problem (Rittel)

According to Rittel and Webber (1973), a wicked problem is a type of problem whose solution unifies the different needs and perspectives of diverse stakeholders. You can only "tame" this type of problem but – per definition – a wicked problem is insoluble.

> "For a wicked Problem there is no definite response" (Rittel and Webber 1973).

Rittel and Webber formulate ten criteria to differentiate and define the nature of wicked problems. To deepen the understanding for wicked problems we will present three criteria.

> "Coexistence of problem and solution" (Rittel and Webber 1973).

Rittel and Webber mention that the problem changes while working on the project and while bringing the problem into question. All imaginable expertise for manifold implementation and scenarios must be checked to generate argumentative-based variance. Scenarios must be developed and checked against each other until one can be considered as the most viable, sustainable and superior one (from a users perspective). To check the internalized assumption of a scenario, prototypes will be built. These prototypes will create resonance and feedback while being tested by users and this contributes to the further working process. For the sake of convenience we labeled the ability to deal with wicked problems and to generate fitting and innovative solutions that contribute something characteristic to the world (from a user perspective). We call it wicked problem solving competence.

3.2 Design Thinking Education Addresses Wicked Problems

The Schools of Design Thinking in Stanford and Potsdam offer a process and a toolbox for methods and techniques (see in this book, previous article v. Thienen et al. 2012/13).

One is, for example, the reframing, the modification of a problem representation after generating new information and insights. Rittel and Webber stress that:

> "The information needed to understand the problem depends upon one's idea for solving it. This is to say: in order to describe a wicked problem in sufficient detail, one has to develop an exhaustive inventory for all conceivable solutions ahead of time." (Rittel and Webber).

The ability to reframe a problem representation several times during the process is central for the solution of wicked problems. A further example for the research phase during the design thinking process is to interview users to gain insights and empathy for their needs. The acceptance of the user and stakeholder is also considered important in order to judge if a solution is better or worse.

> "Solutions to wicked problems are not true-or-false, but better or worse." (Rittel and Webber).

As a further content of the toolbox, we consider helpful the typical school of design thinking setting and a structure to enable the students to deal with wicked problems. For the setting this means, for example, work in small teams of five to six students with multidisciplinary backgrounds who will be supervised by at least one coach to work on challenges and wicked problems. The collaborative work makes it easy to learn from each other and due to the open space even to learn from the other teams. Another supportive aspect is an error-friendly environment and explicit maxims such as "Fail early, fail often," which will be executed and internalized as mindsets by the students. Among the other mentioned aspects this leads to an attitude, among others, to prototype very early in the process. By prototyping as soon as the team can test the assumptions, the prototype is internalized When a prototype is tested with users, in case of dissatisfaction the team can dismiss the idea as soon as possible before spending too much work and time developing it. Thus failure will be part of the process and will not be avoided artificially. The teams take risk by trying out risky solutions and by testing them early. At the schools of design thinking students work in an open space which is only divided by moveable bar tables and whiteboards. In this way it is easy to observe other teams while working and to learn, for instance, what to do well in the sense of "if the assumption fits" the prototypes. Furthermore it is easy to get constructive feedback from the other teams. Among manifold methods, techniques, and settings, prototyping is a core technique of the schools of design thinking toolbox.

3.3 Prototyping Is a Highly Efficient Tool for Tackling Wicked Problems (Arrow 2)

A central technique of the schools of design thinking toolbox is prototyping. Prototyping in design thinking is introduced at a very early stage, often hours after handing out a new project (for example in the fast forward exercise). The aim is among others to empty the brain of biases concerning solutions and to be open-minded (again) for other, better solutions. For it is certainly possible to come back to an early prototype later in the process. The advantage of prototypes is that you can have them visibly around in the workspace (Gerber and Carroll 2012). First ideas will be made tangible and can be modified iteratively. For the early phase of the process techniques, such as paper models or rough prototypes, are quickly realized and offer an easy way to make an idea testable. In addition, when the aim of the team is to involve the user more emotionally, the informative role play is chosen. Thereby, the team plays out their idea or shows their service or product within a context. They often stress the problem with the consequences and focus on the needs of the user.

Prototypes have two characteristic roles in the design process, on the one hand prototypes enable interactions and feedback with the user, client or teammate and make the generation of new information and insights possible. Via a prototype a user can give feedback on the general idea, on a detail or a function and also on whether the solution is a "better or worse" one. On the other hand prototyping is a generative tool for designers that will accompany the draft and design process (Gerber and Carroll 2012). At the same time a designer makes progress in the process, the prototypes reflect the progress of the knowledge and assumptions.

3.4 Prototyping and Dealing with Wicked Problems

According to Rittel and Webber, there is not only one definitive solution. There should be manifold solution proposals more or less qualitatively developed where one is better than the other (Rittel and Webber). The better solution is one that is able

> "to improve some characteristic of the world where people live." (Rittel and Webber)

Given that there is no ultimative test for a solution, the only possibility is a confrontation of the user with a prototype and to test the proposed solution and to analyse the resonance and feedback towards the prototype and the idea behind it. Once the prototype, the service or the product is considered by users and stakeholders as desirable, you can assume to have found one of the "better solutions." While prototyping, the comprehension of the problem and the solution deepens and you can speak of the co-evolution of problem–solution (Dorst and Cross 2001). This leads to an iterative reframing of the problem during the process.

The reframing process is closely linked to prototyping because the new frame of the problem is only verifiable and testable via a prototype. The created new framing for the problem will be manifested in a prototype and the "included assumption" can be tested by user. This is the best way to generate feedback when a good solution for a wicked problem is developed.

Furthermore, every prototype is in a way a visible reminder of a current or older assumption and labels the station within the process. With a collection of various prototypes, the advance in the process is seen and that there is movement even in moments of being stuck. The visual representations of working steps helps to gain confidence in the progress being made and is a type of visual feedback (Gerber and Carroll 2012).

What further factors do we have to include in the solving of complex problems? In cognitive-psychology there are relevant factors for prototyping.

Based on Dörner 1995, the human memory is a medium where all psychological processes of a human being occur. The biggest impact for solving design problems is the working memory, because the working memory connects information from the long-term memory with the short-term memory.

Solving complex problems exclusively in the mind is possible only to a certain extent. This is, because of the limitations of the working memory. For thinking, complex problem solving needs enormous resources. Merely to "invoke" a more complex shape in the mind requires capacities and concentration. Once this capacity is committed then it is not possible to go back to the further development of this shape. Here external representations are a big support. The sketching and the three dimensional modeling of a paper model, for example, are supportive of the thinking and again to dispose resources for problem solving (Dörner 1995; Wiese and Wiese 2012). An overload of the working memory in solving problems find expression in a too huge simplification on too few influencing variables of a draft problem (Ehrlenspiel 1993). As a consequence, the problem solver modifies his procedure to fit his limited resources and will work with his simplified representations of the problem and approximate problem-solving strategies (Klauer et al.). It is not possible that complex draft work can be executed in parallel, but rather sequentially. Furthermore, it is preferable to work on already known solutions. This implies that a problem solver who works with an overloaded working memory tends to reduce the complexity of a problem and to solve it in a less complex way than the problem would require. To release the working memory, while solving complex problems. an externalisation of the mental problem representation would be appreciated. This is especially the case when there is a certain increased complexity of the problem as with wicked problems. The cognitive conditions as a free working memory and to deal adequately with these conditions as well as to work with external representations leads to more creative solutions. In parallel to the need to externalise (mental representations) for more creative solutions there is as well the advantage that early prototyping enables time saving, cost saving and better results. As mentioned in the previous chapter, there are, among others, cognitive psychological aspects that makes plausible that prototyping is a technique to solve complex problems. Prototyping is an important factor to support wicked problem solving competence. Furthermore, the belief in one's own successfully experienced ability to solve wicked problems is important to being a self-efficacious and competent wicked problem solver.

4 Wicked Problem Self-Efficacy Is a Precondition of Wicked Problem Solving Competence (Arrow 3)

4.1 Wicked Problem Solving Competence

First of all, a competent wicked problem solver is a person with a human-centered perspective of the world and a human-centered approach to solutions for wicked problems. In addition, it is necessary to have the ability to deal with ambiguity and to generate feedback on ideas via external representations, as is the case via prototypes. Further needed is a sense of openmindedness for an ambivalent process, a feel for when all information is collected, and an ability to decide which solution is better than another one. Design thinking works with a human-centered approach. For example, in a research phase, students learn how to develop empathy for the user and how to learn more about their problems and their needs. More advanced wicked problem solvers don't have a fear of failure and take risks. We label a behavior that leads to a good solution for a wicked problem as wicked problem solving competence.

4.2 A Specific Form of Self-Efficacy: Wicked Problem Self-Efficacy

Bandura developed the construct of self-efficacy, which he defines as the belief in one's own ability in a specific domain:

> "increased feelings of ability influences what goals people set, how much effort is expended, perception of difficulty of engaging in tasks, persistence, and attribution and resilience to failure." (Bandura 1997)

As more self-efficacious people have a higher motivation and a higher tolerance for frustration, it is plausible that the performance and the outcome will be better.

Especially in a context of wicked problems, high perseverance is needed in a context where complex problems are considered as "wicked" and "mean." Often setbacks and detours are connected to wicked problems and because of the coexistence of problems and solutions there is a continuous query. Important for dealing with wicked problems is the ability to accept in the process ambiguity, fuzziness and a permanent change of certitude and a loss of control. Being self-efficacious in solving wicked problems is addressed by the schools of design thinking toolbox. To stress the importance of wicked problem self-efficacy you may say that you won't start to work on a wicked problem unless you believe you can solve it.

5 Prototyping Is a Highly Efficient Tool to Foster Wicked Problem Self-Efficacy (Arrow 4)

When the right prototype was chosen, the adequate technique was to incorporate the idea and to test it with users. When the user "understands" the question that the prototype asks, the prototyper has a feeling of success, even if the user doesn't like the outcome. Another positive experience of prototyping is when a prototype is tested by a user or stakeholder and the resonance makes it possible to decide that a solution fits a user's vision and is a good solution for the wicked problem. This experience leads to an enhanced wicked problem self-efficacy.

The successful prototyper receives confidence and a feeling of control in a highly ambiguous process. According to Bandura, experiences of success lead to mastery experience which lead to the fact that more resources will be activated to achieve a goal, so even if there are setbacks more perseverance is executed.

> "Individuals are more likely to put forth greater effort and commitment to a given task than they are to self-doubts (Bandura 1997). "

The user acceptance of a prototype and the included solution proposal creates positive experiences and an enhanced self-efficacy. Students realize that their prototyping skills are a reliable and good tool to realize and recognize complex solutions for wicked problems.

> "Perceptions of ability to control are developed through mastery experiences (Bandura 1997)."

Mastery experience is one of four sources of self-efficacy among social learning, verbal persuasion, psychological and affective states, as described by Bandura. The impact of prototyping for an enhanced self-efficacy is researched by Gerber and Carroll (2012). The authors examined in a high-tech firm the psychological experience of engaging in the practice of low-fidelity prototyping. "The study finds that the production and rapid visualization of multiple ideas through low-fidelity prototyping allows practitioners to reframe failure as an opportunity for learning, support a sense of forward progress, and strengthens beliefs about creative ability." Self-efficacy is enhanced as well through feedback (social learning). At the schools of design thinking there is a vital culture of feedback. For example there is the maxim of "defer judgment". When giving feedback the rule is to express feedback in a constructive way and in form of I like and I wish. The culture of feedback is made visible by a ritualized daily session in the end of the day, the so called, I like I wish session. There is only constructive feedback allowed and a "I wish feedback" goes normally with a concrete "how to". So feedback and setbacks don't discourage you but show you new ways and perspectives on your work and motivate you to attack even tasks considered as real challenges or too difficult to even try to start with. Bandura mentioned that

> "Fear of failure reduces overtime, allowing individuals to take on more ambitious challenges (Bandura 1997)."

Prototyping seems to be important for the self-efficacy of the problem solver and the ability to develop manifold and assumption-adequate prototypes is addressed by the schools of design thinking toolbox.

6 Outlook

As mentioned in the chapter above, there are aspects in cognition psychology, which make plausible that prototyping is a technique to foster wicked problem solving competence. Prototyping is an important factor to enhance wicked problem solving competence, as well as the belief in one's own ability and experiences as a good wicked problem solver. The proposed model shows our assumptions that the interactions of the various factors are plausible, while at the same time the model is not the ultimate truth. To be able to test our assumptions and this model we developed prototyping technique cards to give orientation in choosing the fitting prototyping technique for design thinking novices. During the Basic Track at D-School in Potsdam we plan to have a test group who will work with the developed set of cards regularly, a test group who will work with them at rare intervals, and one group who will work without the cards. The D-School teams who will work with the cards will be observed when using them by video. We assume that the cards will have effects. For example, when the design thinker novice is using the cards, there will be an improvement in her prototyping skills and the wicked problem self-efficacy. Her wicked problem-solving competence will also increase. To measure these effects we developed a questionnaire building on the ten criteria of wicked problems by Rittel and Webber. By analysing the video data and by handing out the questionnaires in a pre and post-test setting, we will get data to gain insight on whether the prototyping skills or the selection of adequate prototypes is be improved by using the cards. At the same time, we will see if there is an enhancement in wicked problem self-efficacy or wicked problem solving competence during the design thinking education at D-School Potsdam.

References

Bandura A (1997) Self-efficacy: the exercise of control. W.H. Freeman, New York

Buchanan R (1992) Wicked problems in design thinking. Design issues 8(2):5–21

Dörner D (1995) Problemlösen und Gedächtnis. In: Gedächtnis D (ed) Probleme-Trends-Perspektiven. Hogrefe, Göttingen/Bern/Toronto/Seattle, pp 295–320

Dorst K, Cross N (2001) Creativity in the design process: co-evolution of problem–solution. Design Stud 22:425–437

Ehrlenspiel K (1993) Denkfehler bei der Maschinenkonstruktion. In: Ja, mach nur einen Plan! Huber, Bern, pp 196–207

Gerber E, Carroll M (2012) The psychological experience of prototyping. Design Stud 33(1): 64–84

Giddens A (1996) Konsequenzen der Moderne, 7th edn. Suhrkamp Verlag

Klauer KC, Hrsg. B. Strauß und Kleinmann M Grundlagen der Problemlöseforschung. In: Computersimulierte Szenarien in der Personalarbeit. Verlag für Angewandte Psychologie, Göttingen, pp 17–42

Lindberg T, Raja G, Birgit J, Christoph M (2010) Is there a need for a design thinking process? In: Proceedings of design thinking research symposium 8 (Design 2010), Sydney

Rittel HWJ, Webber MM (1973) Dilemmas in a general theory of planning. Policy Sci 4(2): 155–169

Wiese E, Wiese L (2012) Designproblemlösen mit externen Repräsentationen. In: Prototyping! Physical, virtual, hybrid, smart. Form + Zweck Verlag, Berlin, pp 160–185

Creative Collaboration in Real World Settings

Matthias Wenzel, Lutz Gericke, Raja Gumienny, and Christoph Meinel

Abstract Tele-Board was designed and implemented for use in Design Thinking teams. From the development of simple prototypes, we came up with a reliable software system for a wide range of users. There is a growing interest in the system by companies who seek to ease their day-to-day collaboration activities worldwide. In this article, we present the results of a study, which we conducted with a company and its implications on the development and adjustments on the system. With the help of usage data, interviews, and observations, we could study the requirements of a global company. It shows how important the flexibility of our system is for a multitude of use cases. Different levels of knowledge, room configurations, as well as different hardware configurations are possible and well supported. The feedback and statistics we got from the study are evidence of a large level of acceptance and successful adoption by the users. We reflect and abstract the lessons we learned from that specific user group into future developments and research opportunities.

1 From Research Prototypes to Real-World Usage

We designed and implemented the Tele-Board system in order to allow Design Thinking at geographically distributed locations. Tele-Board combines the traditional whiteboard and sticky note metaphor with all possibilities of the digital world. The system enables Design Thinkers to work as they are used to, but also see what their colleagues at other locations are doing or have done before.

During the whole development process we tested the system with users and, especially in the latest states, we deployed Tele-Board in different situations and student teams (Gumienny et al. 2012).

M. Wenzel (✉) • L. Gericke • R. Gumienny • C. Meinel
Hasso-Plattner-Institute, University of Potsdam, 14482 Potsdam, Germany
e-mail: tele-board@hpi.uni-potsdam.de

H. Plattner et al. (eds.), *Design Thinking Research*, Understanding Innovation,
DOI 10.1007/978-3-319-01303-9_9,

Though we learned a lot through the studies with student teams and use in different contexts, we now deployed Tele-Board for a longer time period and also in an industry setting. In this setup, we could make use of Tele-Board's advancements in former research and had the chance to equip industry partners with the system. This way, we conducted long-term research in a real working environment of a digital whiteboard system that has not been possible before. In the most comprehensive study so far, we analyzed the system log of a 3 months usage period of a team working in a large IT company. The team members are located in Germany, Italy and India. In addition to the system log data, we also interviewed the users about their experience with Tele-Board and how it had changed their collaboration and communication behavior.

Here, we will now give a detailed description of the study and the insights we gained. We will show which functions and properties of the system foster remote collaboration and what could be adjusted and optimized for users in a large company.

2 Introduction to Tele-Board

In order to understand the design and setup of Tele-Board and how it was used during the study, we will first give a short introduction of the Tele-Board system and how it has developed over the last years. For more details on its architecture, please see our former publications (Gericke et al. 2012; Gumienny et al. 2011).

Tele-Board is a digital whiteboard system supporting creative teamwork, such as design thinking, especially if team members are distributed over different locations. All users share a digital whiteboard surface, where they can draw, create and rearrange sticky notes, change their color, or group them in clusters that can also be moved. All actions are synchronized to each connected whiteboard enabling users to work on the same content simultaneously. Furthermore, users' actions are automatically stored while working on the whiteboard, which allows recreation of a whiteboard's state from any point during its course of development.

2.1 Tele-Board Components

The Tele-Board software system consists of several components. These are: the web portal, whiteboard client, digital sticky note pads, and server component, which run on a range of hardware devices (for an example see Fig. 1).

2.1.1 Web Portal

The web portal acts as the starting point for the Tele-Board software system. Here, users can manage projects and panels in order to organize their work. Within a

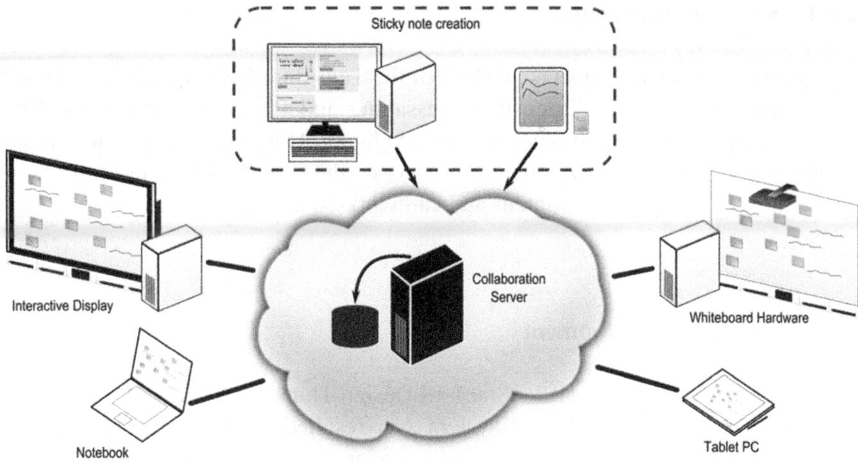

Fig. 1 The Tele-Board software system architecture. All connected devices are synchronized via the central collaboration server. The whiteboard client software can be run on a wide range of hardware devices, which connect to the server component. Sticky notes can be created with the help of mobile devices acting as digital sticky note pads (e.g. tablets or smartphones) or in the web portal

project, users can create various panels, which represent virtual whiteboard spaces. Editing of a panel is facilitated by the whiteboard client, which is started directly from the web portal. This way, installation of any additional software is not necessary for using Tele-Board.

2.1.2 Whiteboard Client

The Tele-Board whiteboard client is the digital counterpart of a traditional whiteboard. It is a Java application and can therefore be opened on standard computers. The whiteboard client is independent of the preferred input devices i.e. it can be operated with whiteboard hardware connected to a user's computer, a tablet PC or just with a mouse on the desktop screen. Every interaction with the content (e.g. drawing, moving sticky notes) is transferred to the Tele-Board server component, which ensures the synchronization of all connected whiteboard clients.

2.1.3 Digital Sticky Note Pad

As a digital equivalent for paper sticky notes, different applications for writing sticky notes on mobile devices are provided (see Fig. 1). The Java application is best suited for tablet PCs and other pen-based input devices. Moreover, we developed apps for iOS and Android devices, which can be downloaded from the respective app stores free of charge. Users can create a sticky note on one of these devices and send to the whiteboard afterwards.

2.1.4 Server Component

The server component connects all parts of the Tele-Board system (see Fig. 1). All events are transferred as Extensible Messaging and Presence Protocol (XMPP) messages and distributed to other connected clients. This keeps whiteboard clients synchronized. That is to say, all users, who open the whiteboard client of the same panel, always see the same content and can work on the same items.

2.1.5 Course of Development

Within the last years of the HPI – Stanford Design Thinking Research Program, the development of the Tele-Board system has resulted in a variety of different functionalities. In Fig. 2 we outline the timeline of the Tele-Board development from the beginning in October 2008.

In the beginning, we created an environment for teams applying creative methods that allow them to work together efficiently across distances, without having to change their working modes. In order to deploy a running prototype in a short time, the first version of the whiteboard client was developed using JavaFX. This way, it was possible to gain experience and gather feedback at an early stage.

As we learned from this prototype, it is not only important to enable synchronous working modes for distributed design teams, but asynchronous collaborative work as well. To address the problems of Design Thinking teams, who are working asynchronously over distances, we developed the Tele-Board history browser: a web-based interface making it possible to go back and forth in the timeline of a whiteboard. It enables the design thinker to view the collected data from different perspectives and thereby gain a deeper understanding of the project context. Additionally, it supports teams in analyzing the overall project progress and decision paths taken by the respective distributed sub team or by the team itself in an earlier project phase. The team can also continue at any past state by duplicating the whiteboard content, i.e. starting a parallel session. All data is persisted implicitly, meaning that the user has the freedom not to think about saving data. The JavaFX technology turned out – at that point in time – to not be fast and flexible enough. Therefore, we decided to develop a Java Swing based version of the whiteboard client in order to have more control of the rendering.

Based on insights we got from various user tests, our system was further refined mostly concerning the way users interact with the system e.g. editing of sticky notes and clustering. Moreover, to make asynchronous working modes more convenient and enable reuse of the produced data, we wanted to find ways of better documenting and computationally "understand" the content. Based on the data that is already archived in our system, we evaluated different approaches in the field of handwriting recognition and their applicability for unstructured whiteboard notes. Applications for the recognized texts are searches on the whiteboard content or reuse in text-applications, e.g. spreadsheets or presentations.

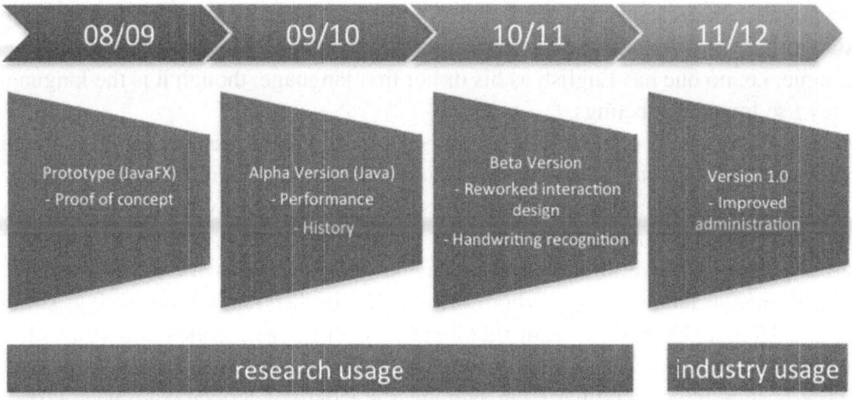

Fig. 2 Timeline of the development of Tele-Board

Towards more fine-grained administration functionality, we released Tele-Board version 1.0 in early 2012. With the current system it is possible to support creative teams for their synchronous and asynchronous work in academic as well as in industry contexts. Additionally, our all-digital tool presents opportunities for numerous functions that cannot be applied in analog environments. The latest released version is being used in a corporate environment for the first time in the current year. Our experiences and findings of how Tele-Board was used in an industry context are presented in the next sections.

3 Real World Usage Scenarios

Collaborative work over distances is common practice for most employees of large global enterprises (Koehne et al. 2012). Many meetings include participants from multiple locations worldwide (Espinosa and Pickering 2006; Hinds and McGrath 2006). Although a variety of tools for supporting different working modes exist, the most commonly used tools are still audio conferences (for synchronous communication) and e-mail (for asynchronous communication) because they are readily available and easy to use (Matthews et al. 2011; Tang et al. 2011).

Still, some ways of working, such as collaborative brainstorming or sketching ideas on a whiteboard, cannot be completely supported in remote situations using standard tools. But this way of working is becoming more and more prevalent. It is even seen in larger companies who seek to introduce methods such as design thinking (Brown 2009) in order to increase their innovative potential (Martin 2009).

3.1 Challenges of Remote Teamwork

A lot of teams in large global enterprises are spread throughout different locations and have to work together remotely with access only to an audio conference and their computers for most of their meetings. For this study, we worked with a team of

25 people, where 15 of them are located at the company's headquarter in Germany, two in Italy and eight in India. All of them have the local language as their mother tongue, i.e. no one has English as his or her first language, though it is the language they use for their meetings.

As part of an IT company, the team's job is to create pre-configured software packages that are ready to use by the customer. When organizational topics and new ideas are discussed in their team meetings they meet bi-weekly with all members of the team and use a video-conference for communication. All team members are working on several software packages, also together with different people outside of the team some of whom are located in different countries, e.g. China or the United States. The team members as well as their colleagues from other teams have a variety of different academic backgrounds, that is to say their team work is quite interdisciplinary. On average, the length of service with the company is higher with the German colleagues (on average 13.5 years) than the Indian (avg. 6.3 years) and Italian (avg. 8.5 years) colleagues. Because of the different expertise and new members in the team, knowledge transfer across locations is necessary.

Two months before the beginning of the study, we conducted interviews with six people from the team in order to identify their challenges and needs.

Over 60 % of the team's work is collaborative with other colleagues within the company. From collaboration on past and present projects, everyone is used to working with remote colleagues. And, as also reported by other researchers (Espinosa and Pickering 2006; Tang et al. 2011), time zone differences are a "*major pain point*" for collaborative work. For their current team, they see it as an advantage that just two time zones with a time difference of less than 5 h are covered. Still, communication with other locations is not always easy. Although two interview partners said that they can clarify some issues over the phone or via instant messengers, others saw a hurdle in just calling their remote colleagues:

> Perhaps it's the distance and the time difference, but you don't just pick up the phone and say: I have to tell you an idea I just had. That just doesn't work. I mean, you don't want to waste someone's time with any vague idea.

With their co-located colleagues they would just walk by the office and see if they had time. Another difficulty for the communication with other locations is language barriers. As already stated above, no one has English as the first language and speaks a specific accent, which is sometimes hard to understand by the others and can lead to misunderstandings. But not only understanding is difficult, but the same is true for formulating ideas and thoughts in another language:

> Ideas or something you want to do spontaneously, in another language you have to think: how do I phrase this? And this is not only our problem that we have to speak English, it's the same for people in India and Italy. Nobody can speak their first language and I think this is indeed a little handicap.

Especially the topic of idea generation and collecting the points of view from people with diverse backgrounds and countries is seen as a great advantage of their global team. They told us, that their manager some time ago, had started an exercise where everybody was supposed to come up with challenges and opportunities for a specific topic. The exercise was done simultaneously via all three locations and was perceived differently by the participants as e.g. these two interviewees:

> You have to use all the communication channels: mails, phone, audio conference options and you have to collaborate, ideate, come up with your thoughts, collate. And then present back in the same meeting. So it's both interesting and challenging.
>
> My experience was: It was a rather chaotic situation. We were all sitting in one room and then split up into different groups. One person was on the phone and the others were brainstorming from the side, and I was typing in everything in a word document. Others came up with using Excel... so the approaches were totally different.

The mentioned tools, such as email, Word and Excel, are indeed used for remote collaboration, but they are not ideal for idea generation or brainstorming. We also heard that people use desktop sharing for the majority of their meetings in order to show PowerPoint slides or other things on their desktops. However, they sometimes apply other ways of working that are not supported by the current tools:

> We had a meeting two weeks ago with another colleague in Germany. While one was drawing something on a whiteboard, the person who was not in the room was lost because he could not follow the discussion in that sense.

Above, we outlined the most important challenges of distributed work for the team members we interviewed. Summing up, they have to find a way to collaborate more closely, even if their colleagues are far away, their office hours are not the same and everybody has to communicate in a foreign language. Additionally, they wish for more support for new ways of working such as idea collection with the whole team or working with whiteboards or other tools that are currently only available for co-located teams.

3.2 *Evaluation of the Real World Study*

Tele-Board was introduced to the team in a workshop at the beginning of February 2012. All of the functions were shown in detail and the participants could practice using them. Afterwards, the team started to use Tele-Board during their daily work. In the first 3 weeks, only three to four people used the systems, but in the following weeks more and more users were involved and the whiteboard usage increased. During the Easter vacation time in Europe the number of users dropped, but afterwards we saw a continuous increase of user involvement and whiteboard usage.

Time zone differences can be a challenge for teams at different locations. But as Tele-Board also supports asynchronous ways of working, we were interested in the distribution of working hours. Figure 3 shows the aggregated number of users for every hour of the day for the entire 12-week period. For example, we can see that all users in India had worked with the system between 12:30 p.m. and 4:30 p.m. local time at some point during the study period. Between 8:30 a.m. and 9:30 a.m. at least half of the users connected to the system one or more times.

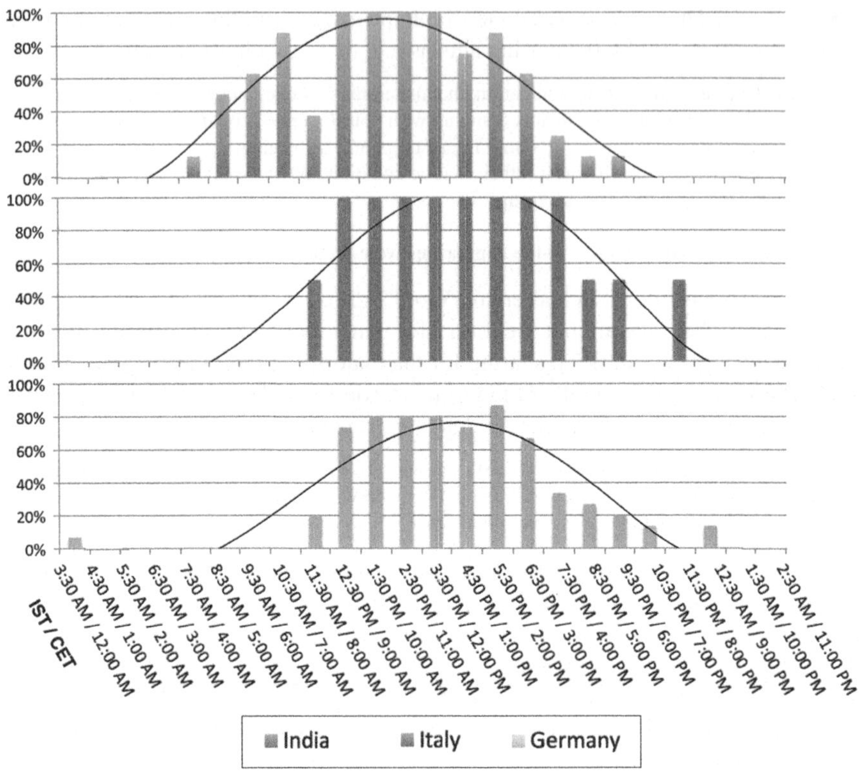

Fig. 3 Distribution of user involvement throughout the hours of the day and over the whole study period. Locations are differentiated by colors; the two involved time zones are Central European Time (CET) and Indian Standard Time (IST)

The team mostly followed the typical working hours pattern from about 9 a.m. to 5 p.m. local time. Apparently, the overlapping synchronous work time is more intensively used than the time spent working alone at one-location only. This is the case, even if it implies that the Indian colleagues are working until 6:30 p.m. or longer. In upcoming research, we want to investigate if this intense synchronous use can also be observed in teams that have fewer overlapping working hours. We expect to see more asynchronous work because it is likely that people prefer working during the standard office hours at their time zone, as we can partly see with the Indian team members here.

Figure 4 shows a selection of panels and how they were used from week 4 until the end of the study period. The size of the circles indicates how many users were involved each day and the colors show the location of the respective users.

The interactions on the panels reveal different usage patterns. It can be assumed, for example, that panel 590 is used for a long-term feedback collection activity. Panel 589 was primarily used as an idea sketchpad during a meeting on that specific

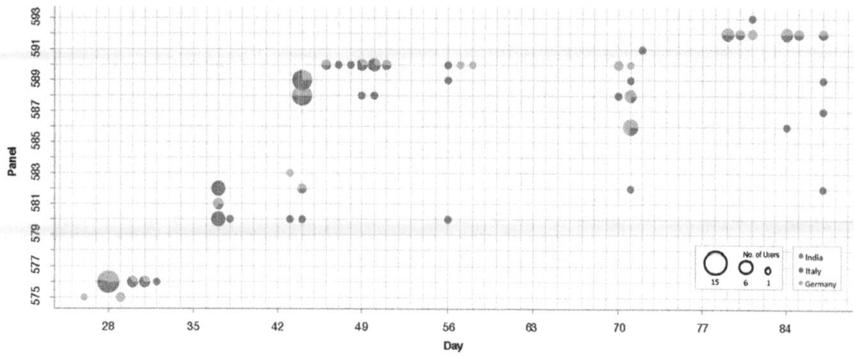

Fig. 4 A selection of panels and their user involvement structure over time. Some panels were only used in a few synchronous global team meetings, while other panels were only used by a few participants asynchronously

day. After this day, it only serves to document meeting recaps (see the interview section as the basis for our assumptions). Looking at the different panels, we can see that there is not one way of working with Tele-Board, people rather shift between working alone, in small groups, and in larger groups. They use the panels during different time periods and manipulate the content asynchronously as well as synchronously.

Giving users the possibility for easy transitions between working together and alone in order to support "loose and tight coupling", which is important for distributed groupware (Gutwin and Greenberg 2002). Our users also appreciated that Tele-Board not only provides one way of working but several that complement each other.

With regard to the degree of collaboration among Tele-Board users, we were interested in how many people worked with the same content, i.e. the same sticky notes. It turned out that almost half of all sticky notes created during the study period were moved/modified by at least 2 people over the course of the study, with the highest number of users editing even eight sticky notes (average: 2.08, SD: 1.44).

The most interesting fact we could derive from the log data was that the activity of all Tele-Board users was not evenly distributed over the different locations (see Fig. 5). For all interactions with the system – number of sticky notes created, number of whiteboard events, number of whiteboard client sessions, and session duration – we could find higher values for users at the company's subsidiaries in India and Italy. Although there were some very active users at the headquarters in Germany, others contributed little or nothing. However, four of the five new users who joined the team after half of the study period are located in Germany.

Because the length of service with the company in Germany is higher on average, we were wondering if there is a correlation between the length of service and the activity of the users (see Fig. 6). In this figure, the size of the circles shows the number of sticky notes a user created (each circle represents a user). The y-axis

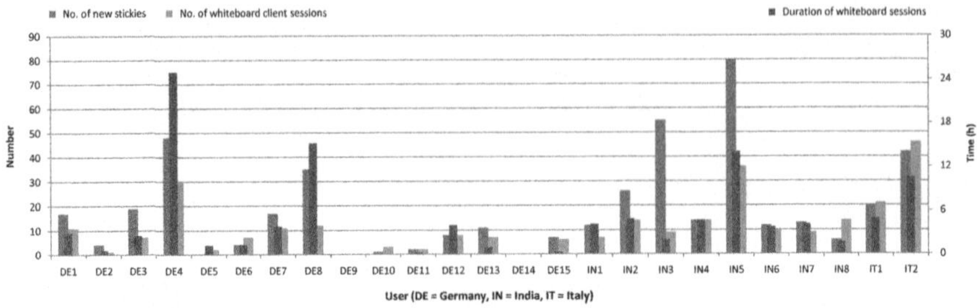

Fig. 5 Activity of all Tele-Board users. The left y-axis indicates the number of whiteboard client sessions (*orange*) and the number of created sticky notes (*blue*) per user. The right y-axis and the *red bars* show how many hours each participant used Tele-Board within the 3 months of our study

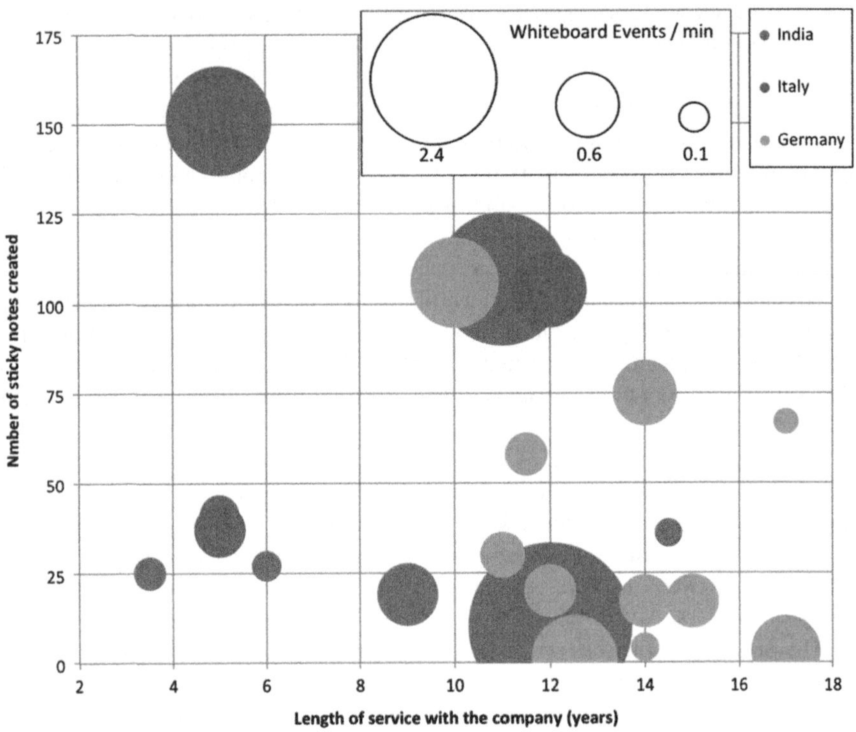

Fig. 6 The level of activity of all Tele-Board users opposed to their length of service with the company. Each user is represented as one circle. The number of sticky notes created is expressed by the size of the circles. Based on their location on the y-axis, the circles also show whiteboard activity in terms of whiteboard events/session duration

indicates the ratio between whiteboard events (= every interaction with the whiteboard client) and session duration, i.e. whiteboard events per minute. This shows how active the users were, after they had opened the whiteboard client. Please note that three German users do not appear in this figure because they did not write any sticky notes. Though the statistical values in the table show that the German users generated fewer whiteboard events, opened the whiteboard client less often and wrote fewer sticky notes (as we can also see here), a lot of them used the whiteboard actively once they had opened it. This is indicated by the ratio of whiteboard events per minute. However, we cannot see a correlation between activity and length of service with the company.

There is a variation in Tele-Board usage between locations. We see the reason for that based on the difference of potential benefits, as described by Gutwin & Greenberg's study on informal collaboration (Gutwin et al. 2008). As mentioned before, there are many domain experts at the headquarters and the German users can just informally meet them on the corridor or go to the other person's office if they have a question. Therefore, the benefit of using Tele-Board at the headquarters is not as high as it is at the subsidiaries and maybe not "*worth the effort of initiating and joining into a collaborative session*" (Gutwin et al. 2008). Especially with regard to knowledge exchange, users at the subsidiaries see Tele-Board as a place where they can ask questions and discuss them. The user saves the time normally spent writing e-mails by implementing Tele-Board instead as the tool for this communication. The difference in activity between locations correlates with the opinions on the usefulness and effectiveness of Tele-Board, as the analysis of the interviews in the next section shows.

4 Interviews with Tele-Board Users

We conducted semi-structured interviews with 20 members of the team. Eleven interviewees came from Germany, seven from India and two from Italy. As four interviewees were new to the team, they did not really use Tele-Board yet, but talked about their first impressions and experiences based on former collaborative remote work. Interview questions focused on the usage of Tele-Board (how and in which ways of working) and the general communication within the team and stakeholders outside of the team.

When we asked how they used Tele-Board, the interviewees said that they use it primarily at their team meetings in the phases of generating ideas (e.g. brainstorming) and collecting feedback. All users created the sticky notes in the same way: in the web portal by typing the note's content with their keyboard. Especially for idea generation, all users like Tele-Board because it "*just works like a whiteboard, a virtual one, as it is intended.*" For synchronous ways of working, the team members typically do a silent brainstorming and afterwards discuss and group their ideas:

> Normally we have a topic for the meeting, and we agree with the participants to allow some time to post the ideas on the board and then we go back into an all-together mode. Normally these are remote, of course, so we are not sitting in the same room anyway. Then we rearrange the ideas on the board and we start sharing and commenting and working on the ideas that are already posted on the board.

Ten out of the sixteen active users told us that they also use Tele-Board for asynchronous working methods. One of them for example, posted his ideas on a whiteboard panel and afterwards he presented them in a project team meeting to the other stakeholders. In most other cases, a meeting organizer creates a panel and sends the link to it together with the meeting request and asks all participants to post agenda topics or questions on sticky notes. In the meeting, they open the panel and go through all topics on the notes. For feedback collection it may happen the other way around, i.e. during a meeting, someone creates a panel and all participants are asked to post their feedback after the meeting (see Fig. 4). This transition between different ways of working is not available in other tools that the team is using and was greatly appreciated.

We just saw that the team uses Tele-Board for idea generation and different ways of communicating synchronously and asynchronously. But, as in the case with all new tools a company implements, the real question is how efficiently the tool supports employees in their daily work. Though we could find different points of view, half of the interviewees stated that Tele-Board saves them time because more people can work together simultaneously. Because of this, meeting minutes and documentation can be omitted and the e-mail correspondence can be decreased.

> Earlier we used to just have an open discussion where anyone who has an idea on a particular topic, gives his or her ideas, and the minutes are taken by one person and finally at the end of the meeting all ideas discussed are sent out as minutes by this person; which is time consuming, it's additional effort for a person to capture the ideas, put them in minutes and send it out. And not all of them speak up during a meeting if they have some ideas. Some of them tell their ideas, some of them don't.
>
> Now that we are extensively using Tele-Board for idea collection, I find that the time for those exercises has drastically been reduced. Because everyone just puts in their ideas and it just takes 2 min. and then it's already there. Now the person who is hosting the meeting only needs to collect the ideas and put it in a proper grouping. The tool has improved the idea collection phase.

Interestingly, we only heard these statements from two people who are located at the company's headquarters. Additionally, eight out of the nine interviewees who work in the other subsidiaries said that Tele-Board saves time. This shows that the benefit of Tele-Board is considered higher by the Italian and Indian users and correlates with the usage statistics of more activity at these locations (see Figs. 5 and 6).

Most interviewees agreed that there are situations when it is not efficient or helpful to use Tele-Board. Such situations include one-on-one phone calls, very short meetings where to-dos are discussed, as well as document review and project status meetings. As the tasks of all team members within their project work are very diverse, some interviewees told us that they rarely have meetings where they can "think out of the box" and the use of Tele-Board would make sense.

Conversely, other users see no advantage of Tele-Board when compared to existing tools:

> I don't see the added value. Basically it's another tool for making notes. I prefer having a list and all of these sticky notes are just too much information for me if no one groups or categorizes them but just sticks them there. You always need someone who sorts them. I prefer having it structured. I prefer a list that I can work from. Otherwise I have to read everything first, then structure it and that's cumbersome.

In general, there are different points of view as to how much Tele-Board can and should help in structuring a meeting. Some think it helps to structure a brainstorming and the grouping of ideas while another user criticizes what he sees as the unstructured format of the meeting and yet another thinks it is good to have a relatively flexible way of working.

Some users especially saw advantages in the asynchronous way of working as time zone differences can be bridged and it is easy to go on working at a whiteboard panel at any point in time, see Fig. 3.

> Sometimes it helps when you are working with colleagues from other locations and you have been doing some tasks and they have to follow up, because of the time difference it's always better you use the tool and you post what you have completed, so you don't wait. Because otherwise we have to wait until the German colleagues come in the afternoon. Here, we can just start in the morning, based on what the German colleagues or others have posted. That might help in going faster.

Another point where users told us that Tele-Board saves some time and effort is with respect to the "clean desktop" policy of the company: all employees are supposed to always wipe off whiteboard content or take away flip-chart sheets and this is not necessary with a virtual whiteboard.

Though some interviewees already perceived the communication within the team as very good, others saw some further improvements with the new tool. For example, one user found it is advantageous for quieter people. The possibility to communicate is easier when it involves posting a sticky note and explaining it afterwards. In general, several people agreed that Tele-Board encourages communication because people are more comfortable to speak freely: They lose their inhibitions to say something if they can post it first. Especially for asynchronous feedback rounds, participants liked the "anonymous" appearance of the keyboard sticky notes though it is indeed possible to track the author of the sticky note. Interestingly, the feeling of anonymity praised by some was viewed by one participant as potentially inhibiting users from posting.

Having everybody's input in a *written* format is important to many users in order not to forget something and to improve the understanding of what people want to say. As stated at the beginning of this paper, problems with audio connections and different accents of non-native speakers sometimes hamper communication. In this case, the written sticky notes can assist in communication:

> When we talk over the phone, it might be that their voice is not clear and we could not understand them properly. In that case Tele-Board was very helpful because we have a written format and we get to see what they really want to tell us.

Overall, the manager of the team also sees an improvement in mutual understanding:

> The understanding is definitely better than before. I mean, I see their results. And when I talk to them separately at regularly scheduled meetings I can tell that they are talking less about different things. The big picture is more consistent. It's not perfect, but much better.

Tele-Board offers a new way of working and the whiteboard and sticky note metaphor is different than the other digital tools used in the company. It was our goal to introduce a virtual whiteboard for remote collaboration which should be, as one user describes it, "*a perfect addition to the other tools*".

As the team formerly used MS Excel for collecting ideas and feedback, some interviewees compared Tele-Board to this tool. One user even had the feeling that they used it like Excel, just the clustering she considered to be easier. The team saw the main difference and advantage as the possibility to enter data simultaneously or "*real-time*." Others thought it was easier to use than other tools because everyone knows the analog equivalent and it is more fun to use because of its colors and playful character.

> [..] it's very receptive, because it's colorful, it's sorted, you can concentrate on the visuals and that's easier to remember than words on an Excel sheet. We did it with Excel before and the rows and columns don't stick in your head. But if you remember the colorful stickies you can say: yes, the pink topic was below the orange one, it stays in your head as a picture.

No matter if people liked working with whiteboards and sticky notes, we often heard that they always have to transfer the content into a MS office document at some point. Or as the manager put it: "*It's good for collecting ideas as a first step towards a solution. But it has to go on…*". This could be a mind map where it is easy to add links and documents or, as some participants suggested, a Powerpoint slide deck.

An additional question to participants was how much they used video conferencing for their remote work. Most of them said that they usually do not use it for their project work, because it is difficult to get a video conferencing room at each location. For this reason, especially in the case of small meetings, video conferencing was seen as not worth the effort. But they also said, it is not very important to see the faces if they see the same content on the screen. One user even thought it could have a negative effect during brainstorming because you are distracted by the faces, which is consistent with other research (Brubaker et al. 2012).

Several users saw it as an advantage that the use of Tele-Board is not difficult to learn and that new users can directly start working after some features are explained to them. These features are: creating sticky notes and adding them to the board, changing the color of sticky notes, creating clusters, and maybe how to create a panel and a project. We heard from the new team members that it was not too difficult to learn how to use Tele-Board.

Members of the team think they would get feedback from other stakeholders (e.g. product owners or consultants) more easily and on a regular basis if these stakeholders had a Tele-Board account.

As the tool was only to be introduced to team members, no automatic routine for creating other users was set up. However, Tele-Board could certainly help in collaborating with people outside of the company as well. As the manager puts it:

> In a world with less budget and possibilities to travel to a customer, a platform for quick exchange is very valuable. It may help to get a faster understanding of the customer's needs.

5 Lessons Learned

While working with different companies, we found out that their specific needs differ from student teams or other groups of people that are not specifically embedded in a very controlled environment. Therefore, we added a lot of features to the system that are more likely to be used in a business setting. Company usage also revealed that some concepts needed to be thought over in a more general way. For example, the management of users and groups as well as the rights assigned to them is crucial in a company. It is also vital to organize and structure the work according to who is working in which team, what are certain dependencies from other teams, and how we can ensure data security in terms of who is allowed to view or edit which content. Bringing all this together, while keeping in mind that Tele-Board should be a tool for creative sessions and easily accessible for almost everybody, was a challenge.

5.1 Optimized, Added, and Re-thought Features

In the following, we outline some of the features that were optimized for company usage.

When we started with the Tele-Board project, we said: We want to make it a tool for everybody who is used to working with a traditional whiteboard. This means virtually everybody. We don't want to have the next graphic editing tool, but we wanted to have it as natural as possible. But determining if the tool feels "natural" is difficult. In the beginning we thought, we would have to transfer everything we know from ordinary whiteboards to the digital world. Copying everything one by one means sticky notes are unchangeable. This limitation was one of the first things we omitted during development. We found that people see, for example, a sticky note text-editing function as a typical tool in the digital world. For a long time we refused to fulfill this requirement, for basically two reasons: First, on a traditional board, you cannot edit stickies at all. Second, we thought that the creation of immutable sticky notes is also part of the design thinking mindset. Going for quantity while deferring judgment – even for what the person herself wrote or said – does not necessarily work together with an editing functionality in the tool. As we have seen that many people are now using Tele-Board on a desktop computer, we decided to allow editing of sticky notes. So far, we have not identified people misusing this functionality and spending too much time on editing single text passages.

There were also some minor changes helping the teams to be more efficient. Starting with an increased number of colors and shades, it becomes easier to differentiate between clusters and sticky notes.

Our observations in larger teams with many simultaneously connected people reflects the need for more synchronicity. In earlier versions, the bin just kept the deleted sticky notes of each single user and people could undo only their own deletion. As there can easily be 25 people logged in, it turned out that it was necessary to also keep the bin global within the session. Still, when the last person closes the panel, the bin is also emptied. This change is consequent in that the state of a session stays global each time.

People use our system while continuing to use their traditional tools. This means we have to integrate our system with other traditional office tools. There are basically two major points of contact for almost every employee in a company today: E-Mail, PowerPoint, and Excel.

As the activity feed is already an important part of the portal system, we decided to also add e-mail notifications so that people are alerted if something has happened on one of their panels and it would make sense to take a closer look.

The integration of PowerPoint was built to ease reuse of the content created using Tele-Board. A whiteboard panel is exported as a slide deck. A cluster on the whiteboard represents a single slide. The topmost sticky within a cluster is assigned as the title of the slide. All other sticky notes are used as bullet points on the slide. To keep readability, multiple slides are created when there are too many sticky notes in a cluster. For having a text-centric output, we also added an Excel export function.

This is the direction from Tele-Board to an external tool, but we also wanted to support the other direction as well – inputting a lot of information as sticky notes to the board. We found that uploading a file using comma-separated values fits most needs. Each row in the file represents a sticky note text. It is also possible to set the color of the sticky using a second attribute. People can use this to input content from a broad range of tools, e.g. spreadsheet applications, word processing, e-mails etc.

Another very important aspect is the involvement of people that are not per se users of the Tele-Board system, which can include any kind of external stakeholders. There are several degrees of involvement: If people just want to take a look at what has been done without editing anything, you can create a public read-only link to a panel. People can use the history browser to see how the content was created. But there is also a second option fostering direct feedback. You can create meetings, allowing also external people to be invited to the session. Using a temporal limited guest access, those invitees can also input into the running sessions by writing sticky notes, while seeing in almost real-time how people are using their input. This possibility was implemented to invite, for instance, external consultants to give their specific expertise during meetings, which can take place on-site or via video or telephone calls.

As mentioned, when developing we had in mind that our solution should be as close as possible to the traditionally used metaphors. We wanted to rebuild the design spaces in the digital world. We accompanied this with a video conference using an overlay approach. This works well in a scenario, where we can install a fixed system in an environment especially furnished for this purpose. But in large companies, meetings are usually handled very flexibly for a range of reasons. Meeting rooms are a shared resource that are booked on demand. Often, people join conferences using their mobile phone or from home. This flexibility cannot guarantee a full-featured video setup in every situation.

The people in our study held a more relaxed view and also saw Tele-Board as a tool that can be used virtually everywhere. So in most cases, people used it on laptops with a mouse and a keyboard. Although we never intended to have a desktop-system, it is a nice serendipity to see that it is flexible enough to be used either way. For many people – especially in our team with employees of an IT company – the keyboard is gradually becoming more "natural" than handwriting. Both modes are well-supported.

6 Discussion and Outlook

In this article we presented a 3 months usage study of a team using a digital whiteboard and sticky note software for their work in a large IT company at three locations on two continents. We have collected quantitative data from usage logs as well as qualitative data from interviews with the users, which allow us to analyze the study procedure from different angles.

From this study, we learned that the "optimal" system setup we created in our lab cannot simply be deployed in a company as it is. It needs to be adjusted to the user's daily environment and equipment, which means standard computers instead of digital whiteboard and sticky note hardware.

However, users can still benefit from a whiteboard software system because it enables them to have real-time idea generation and feedback sessions over distances. It also provides them with a platform for knowledge exchange anytime they want to use it. Tele-Board supports a smooth transition between synchronous and asynchronous ways of working: a user can work at a panel asynchronously and when other users connect to the same panel they are automatically working synchronously. Everyone can choose their preferred way of working, depending on the respective (team) situation. The system log data and interview results show that both ways of working are used by the team. It is therefore important that they are able to easily shift between working alone and working together. Moreover, in the interviews we found that a shared workspace and its artifacts simplify verbal communication and understanding due to the written format of sticky notes. Their graphical note appearance makes sticky notes ideal for arranging and sorting.

6.1 Future Work

A number of computer supported collaborative work systems exist and many new systems come up every day. Acceptable support for teams working remotely and covering *every* aspect of their creative process is still missing. We want to investigate why certain teams perform better than others, which leads us to basically three dimensions of distributed work, we want to research on:

6.1.1 Environment and Degree of Distribution

There are different factors influencing the success of distributed teams. As the communication has to be enabled via digital tools in order to bridge the spatial distances, these tools play an important role in team collaboration. They most certainly have an influence on team spirit, the team's common identity, as well as their general well-being. We want to find out which hardware and software combinations can support these aspects better than others. Typically, there is a combination of telephone, videoconference, and screen sharing equipment applied when working remotely. In a controlled experiment, we want to find out if video-enabled equipment has significant advantages over non-video setups because it supports gestures and facial expression during remote sessions. This could improve the team spirit and common ground and this way enhance the overall experience of remote meetings.

Besides the role of video and audio, we also want to investigate other hardware equipment such as digital whiteboards or devices for creating sticky notes. We want to understand the importance of choosing the right hardware device for a certain situation as well as using it for the right activity. Especially in the interdisciplinary d.school environment, it could be possible that some devices are adopted more easily by all users and others exclude some user groups. Especially for the design of Tele-Board we want to find out which interaction concept works best in order to include all users. In general, we would like to know which factors influence the adoption of different tools and why this is the case.

6.1.2 Support for Different Phases of Creative Work

While investigating the tools used for remote collaboration, it is also important to find out which kind of activity can be supported in which way. The design thinking process model differentiates between six distinct phases (understand, observe, define, ideate, prototype, test). We want to research how far these phases can be supported in remote settings using Tele-Board in the current form and where it has to be changed and in what way. From our experience during the introduction of Tele-Board at a global software company, we know that certain activities are easier to transfer to the digital world than others. We can say Tele-Board is well suited for the ideation phase, which means for applying methods such as brainstorming. Of course, prototyping is much harder to support in a remote setting, because it often makes intensive use of physical material. But still, with the help of videoconferencing some discussion on the prototypes could take place, and there are also other kinds of prototypes, for example screen mockups, that could be created together. As there are different kind of prototypes for different purposes (see the d.incorporate project), it is interesting to find out which of them can be supported in which way. But also for the design thinking phases as e.g. observe or define we want to investigate how they can be supported by different tools, devices and methods. In summary, we want to outline the factors that influence the suitability of a tool to a working mode or activity.

6.1.3 Team Work and Cultural Influences

As the third important column of research, we want to explore the relation of personality traits or general attributes to the successful usage of tools in remote work settings. In earlier research and interviews we conducted with employees of global companies, we identified at least three aspects: the general open(minded) ness of the individual, their experiences and expectations, and the cultural identity.

We use tools that are different from what people are used to at their local workspaces. Accordingly, some degree of willingness and curiosity is required to explore the new possibilities they present – i.e. finding out about the benefits and weaknesses of a tool and learning how to get the most use out of it. We assume this level of impartiality is a key factor. It is influenced by the other dimensions – experience and culture. If we can understand the factors for the impartiality of different users, we can give instructions on how to lower the barrier for certain user groups.

Looking at the design thinking process model, we want to find out, how remote collaboration tools can better support specific phases and activities. We want to create a metric on the influencing factors for successful remote collaboration. Therefore, we will study which design thinking phases can be supported in which way or what is necessary for each phase.

Building up on the lessons learned from the years within the former Tele-Board project and its deployment in academic and corporate environments we also want to find an answer to the question: “What should Tele-Board look like and which features are necessary for an optimal remote design thinking experience?”

With the help of Tele-Board we want to support d.schools in their attempt to collaborate among each other. Other ways of support are the possibility for project partners at different locations to collaborate with the team or the team itself can work together if its members are distributed throughout different locations.

In order to further enhance the system and widen the field of application we want to support companies as they have different requirements than d.schools e.g. limitations concerning hardware or the subsequent processing of generated data.

From the deployment in different working contexts we want to investigate if cultural differences have a significant influence on the usage of our tool.

References

Brown T (2009) Change by design: how design thinking transforms organizations and inspires innovation. Harper Business. New York

Brubaker JR, Gina V, John CT (2012) Focusing on shared experiences: moving beyond the camera in video communication. In Proceedings DIS'12. ACM Press, pp 96–105

Espinosa JA, Pickering C (2006) The effect of time separation on coordination processes and outcomes: a case study. In: Proceedings Hawaii international conference on system sciences (HICSS'06), IEEE, vol 1, IEEE Computer Society, Washington, DC, USA, p 25.2

Gericke L, Gumienny R, Meinel C (2012) Tele-Board: follow the traces of your design process history. In: Plattner H, Meinel C, Leifer L (eds) Design thinking research. Springer, Berlin/Heidelberg, pp 15–29

Gumienny R, Meinel C, Gericke L, Quasthoff M, LoBue P, Willems C (2011) Tele-Board: enabling efficient collaboration in digital design spaces across time and distance. In: Plattner H, Meinel C, Leifer L (eds) Design thinking, understand – improve – apply. Springer, Berlin/Heidelberg, pp 147–164

Gumienny R, Gericke L, Wenzel M, Meinel C (2012) Tele-Board in use: applying a digital whiteboard system in different situations and setups. In: Plattner H, Meinel C, Leifer L (eds) Design thinking research. Springer, Berlin/Heidelberg, pp 109–125

Gutwin C, Greenberg S (2002) A descriptive framework of workspace awareness for real-time groupware. Computer Supported Cooperative Work (CSCW) 11(3):411–446

Gutwin C, Greenberg S, Blum R, Dyck J, Tee K, McEwan G (2008) Supporting informal collaboration in shared-workspace groupware. Journal of Universal Computer Science – JUCS 14(9):1411–1434

Hinds P, McGrath C (2006) Structures that work: social structure, work structure and coordination ease in geographically distributed teams. In: Proceedings CSCW'06, ACM Press, New York, NY, USA, pp 343–352

Koehne B, Shih PC, Olson JS (2012) Remote and alone: coping with being the remote member on the team. In: Proceedings CSCW'12, ACM Press, New York, NY, USA, pp 1257–1266

Martin RL (2009) The design of business: why design thinking is the next competitive advantage. Harvard Business Press, Boston

Matthews T, Steve W, Thomas M, Sandra Y (2011). Collaboration personas: a new approach to designing workplace collaboration tools. In Proceedings CHI'11, ACM Press, New York, NY, USA, pp 2247–2256

Tang, JC, Zhao C, Cao X (2011) Your time zone or mine?: a study of globally time zone-shifted collaboration. In Proceedings CSCW'11, ACM Press, New York, NY, USA, pp 235–244

User-Centered Innovation for the Design and Development of Complex Products and Systems

Lauren Aquino Shluzas, Martin Steinert, and Riitta Katila

Abstract In this chapter, we examine user interaction for the design and development of complex products and systems. Through a two-phase research effort, we explore and test the influence of user involvement (i.e. novice/average and expert/lead users) in early stage design and new product development.

In Phase I, we document the roles of users and stakeholders in early stage design, based on a review of existing literature and inputs from practitioners in the design and engineering fields. From this review, we develop a systematic framework for determining which user entities designers should target for the design and development of new products and systems. In order to increase adoption success, the *Design Stakeholder Identification, Assessment and Ranking Framework* is optimized to specifically address the needs of users and stakeholders with the greatest degree of influence over product use and adoption.

Phase II of this research captures the effects of including expert and novice product users at discrete phases of the product development process, with an emphasis on early stage concept design and ideation. Through a controlled experiment, novice and expert users from the healthcare setting provided inputs for the conceptualization of an intramuscular drug delivery system. Based on inputs from users of each group, four conceptual prototypes were developed. Using quantitative assessment methods, we are currently examining if there are significant differences between concepts that have been designed based on inputs from users of either group. Factors of particular interest include the perceived usability, functionality, efficiency, adaptability, and cost-benefit of product concepts.

L.A. Shluzas, Ph.D. (✉) • M. Steinert, Ph.D.
Center for Design Research, Building 560, 424 Panama Mall, 94305-2232 Stanford, CA, USA
e-mail: lauren.aquino@stanford.edu; steinert@stanford.edu

R. Katila, Ph.D.
Department of Management Science & Engineering, Huang 336, 94305 Stanford, CA, USA
e-mail: rkatila@stanford.edu

H. Plattner et al. (eds.), *Design Thinking Research*, Understanding Innovation,
DOI 10.1007/978-3-319-01303-9_10,

Through Phase I and II of this work, we aim to impact design-thinking research through providing an improved understanding of user and stakeholder roles in the development of complex systems, and new insights regarding user-centric product design teams.

1 Background

One of the core design thinking principles is user-centric design. It not only demands thorough needs finding methods and anthropological approaches, but also requires development teams to test iterative prototypes directly on users, whenever possible. For the design of complex products and systems, identifying the "real" user among multiple stakeholders is a difficult challenge. In the healthcare field, for instance, although patients are often considered the "users" of medical technology, so too are doctors, nurses, hospitals, and insurers. Thus, the question remains: who is the target user? Once "the" user group is identified (if there is only one), then which subjects does one design for within a specific user group – a highly advanced and technologically adept lead/expert user, or the average user?

There is an established field of research on the role of "lead users" that emerged from studies on sources of innovation (von Hippel 1976; von Hippel 1986). Lead users have been described as individuals or organizations who experience needs for a given innovation earlier than the majority of the target market (von Hippel 1986), and who are positioned to benefit from obtaining a solution to those needs. Existing research has shown that users, as opposed to manufacturers or suppliers, have been the first to develop new commercially successful products, with user innovation concentrated among the lead users of those products and processes (von Hippel 1986; Urban and von Hippel 1988; Shah 1999; Lüthje 2003).

However, conflicting views regarding the role of lead users in product design have been conveyed. Ulwick (2002) described that, "lead users can offer product ideas, but since they are not average users, the products that spring from their recommendations may have limited appeal." Similarly, a retrospective analysis of eight medical device firms (Shluzas 2011; Shluzas et al. 2011; Shluzas and Leifer 2012) illustrated that companies often prefer to work with industry thought leaders in the conceptual design and product testing phases of development. However, surgical procedures developed primarily based on input from lead-user surgeons often result in the development of technology that is difficult for a broad range of product users to embrace. To close the usability gap between first and second-generation products, the companies studied implemented post-market revisions based on feedback from "average users" in the clinical field.

In drawing analogies to the Technology Adoption Lifecycle (Moore 1991), lead users often demonstrate similar characteristics to those of innovators and early adopters. That is, they are often risk takers, have a high degree of opinion leadership, and are fast to adopt new innovations (Rogers 1962). In contrast, average users are frequently aligned with the early majority and late majority of consumers in an adoption lifecycle. They are typically more risk averse and tend to adopt new innovations after varying degrees of time (Rogers 1962).

A review of existing literature on user interaction for the development of complex products and systems reveals the need for a greater understanding of the dynamics and inter-relationships among user types across a range of industry settings. In particular, there is a need for research that examines how lead/expert users (e.g. early adopters and innovators) and average/novice product users (e.g. the early majority) uniquely contribute to the design, and subsequent adoption, of products and services.

In this chapter, we present a multi-phase research effort aimed at examining user interaction for the design and development of complex products and systems, across a range of high-technology industries. The first phase of research focuses on the question: Which users and stakeholder groups should developers target for the design of complex products and systems? The second phase of research then centers on the questions: Within each targeted user group, which subjects does one design with and for during the product development process – a technologically adept expert user or the "average" user; and does designing with and for particular users and user groups have an impact on the usability, functionality, novelty, and cost-benefit of new products and systems?

2 Phase I: The Identification of Target Users and Stakeholders

Which users and stakeholder groups should developers target for the design of complex products and systems?

To explore this question, we first present a literature-based comparison between the conceptual notion of "users" and the direct inclusion of "users," within several design methods. After having determined that the concept of "user" varies greatly among design methods and throughout stages of product innovation, we modify the notion of "user" in favor of a more balanced, role-based stakeholder perspective. In doing so, we provide a systematic framework for determining which user entities designers should focus on throughout the design process. Contrary to current processes for assessing and prioritizing users and stakeholders that are often fragmented and qualitative, the framework we propose combines new and existing tools into one systematic process. The *Design Stakeholder Identification, Assessment and Ranking Framework* (Agrawal et al. 2012) is geared toward the design and development of complex systems, involving an embedded network or ecosystem of users and stakeholders. It intends to provide designers with a comprehensive tool for ensuring that the needs of stakeholders with a high degree of influence over product/system adoption are met.

2.1 *The Conflicting Notion of "Users" in Existing User-Centric Design Methods*

The notion of "user" is central to design methods and approaches. According to (von Hippel 1976) 75 % of the commercially successful industrial goods innovation projects were user-driven, rather than technology driven. Depending on the extent of user involvement, there is a continuum of user concepts spanning from a concrete existing user (e.g. lead user); to an observable user or participatory designer with whom one may directly interact (Abras et al. 2004); to an abstract conceptualization of user that views users as customers in axiomatic design (Suh 2001). Since users play a critical role in determining the success of a product and its impact on a community, the selection of the 'right user group' during the design process is critical for the success of any design method. The 'right user group' may vary from one industry to another and from one product to another. In fact there might be more than one 'right user group' for the same product within the same industry.

Every design method has its own interpretation of user and its own criteria for defining the role of the user. As shown in Table 1, we broadly segment design methods into two categories: artistry inspired design methods that focus on iteratively learning from, interacting with, and reflecting on real users; and science inspired design methods that generally have an abstract, conceptual notion of users. The former category, inspired by Donald Schön, focuses on generative metaphors and learning systems (Schön 1983). Design methods derived from this artistry inspired design understanding are more applicable to fuzzy front-end exploration design tasks. The latter category was inspired by the work of Herbert Simon and focuses on quality and efficiency (Simon 1976). Methods from this science inspired design approach are generally more applicable to technical design optimization tasks. For design methods in each category, we present examples of the types of users involved and their corresponding degrees of user involvement.

2.2 *Importance of Selecting the "Right" Users and Stakeholders in the Design Process*

The range of user types shown in Table 1 illustrate that target users may vary for the development of an individual product, based on the design method chosen. For example, the current development of Boeing's 787 Dreamliner has involved multiple users, including the crew, passengers, airline operators, aircraft manufacturers, and regulatory authorities. Although passengers are one of the aircraft's final end users, so too are pilots and the crew. But the choice of buying an aircraft will depend on airline companies. Additionally, maintenance is a major factor in determining the actual airtime for a plane. Thus, involving the 'right user' in the design process is extremely important for the success and adoption of this product.

Table 1 Artistry and science inspired user-centered design methods

Type of design method	Example methods	User types	User involvement
Science inspired design methods: Focus on quality and efficiency; More applicable to technical/scientific design optimization tasks; Promoted e.g. by Simon (1976)	Lean Six sigma process excellence Quality function deployment Cost-benefit analysis Choice modeling Minimum viable product	Users as abstract concept of customers and stakeholders	Limited to no involvement
Artistry inspired design methods: Focus on generative metaphor and learning systems; More applicable to fuzzy front end exploration design tasks; Promoted e.g. by Schön (1983)	Lead user innovation Participatory design Inclusive/universal design Usability/human factors design Experience design Crowd sourcing	Real users, may be separated into adopter categories (innovators to laggards)	Active involvement

For simplicity sake, but also based on existing teaching dogma, users are quite often seen as individuals; when in reality, users are part of a complex stakeholder network with delicate and often non-obvious relationships. Per the 'Stakeholder Network Theory' (Rowley 1997), each firm encounters a different set of stakeholders that aggregate into unique patterns of influence. They respond to the interaction of multiple influences from the entire stakeholder set. In fact, the notion of conducting participatory observations and creating concrete user narratives and personas is gaining ground in design practices. But, there is a danger that designers may focus on the "wrong user," one who may not significantly contribute to the future adoption of the product/service/solution. We therefore aim to identify and analyze the stakeholder network, as a foundation for the user group selection process.

Also, the roles of the user groups may vary significantly from one industry to another. For instance, the government may be the end-user for defense products, but it would be a legislative authority for a healthcare product. Thus, understanding the user in terms of stakeholders and their roles in the context of different industries is critical to developing a framework for classifying user-centric design methods. Due to this context dependency, each product/system and industry requires an independent analysis to identify each user group's specific role in a particular design situation.

2.3 Design Stakeholder Identification, Assessment and Ranking Framework

We present a systematic framework to help design teams better understand the roles of different users and stakeholders, and to prioritize their involvement in the design process. The four stages of the *Design Stakeholder Identification, Assessment and Ranking Framework*, detailed in the paper by Agrawal et al. (2012), include the following steps:

1. Gap analysis (or needs finding)
2. Customer value chain analysis
3a. Cycle of use analysis
3b. Monetary flow analysis
4. Final stakeholder ranking

Step 1 (pre-step) – Gap analysis

The purpose of this initial step is to identify a potential new product, or new features of an existing product, that address unmet needs. This step is intended to ensure that the design team has articulated a clear strategic focus such as a specific industry, target market, or a specific need. There should be a strong match between the expertise in the product team, the industry or market focus, and the potential of the market for product adoption. The outcome of this initial step is a list of requirements for the new product or product features. These could be grouped by functionality or use cases if necessary.

Step 2 – Customer value chain analysis (CVCA)

The purpose of this analysis step is to enable the design team to comprehensively identify pertinent stakeholders, their relationships with each other, and their role in the product's life cycle. The outcome of this analysis is a product value net that includes a list of all product stakeholders, their intents and incentives, and their inter-relationships. Typical value propositions for stakeholders include: money or payments, complaints, regulatory influences, services or necessary information, and tangibles such as hardware and materials.

The CVCA analysis (Donaldson et al. 2006) enables design teams to determine stakeholders with decision-making influence, based on the number of arrows pointing into or away from each stakeholder in the value net. Stakeholders with peripheral influence will less likely be at a focal point in the value net. As such, the value net diagram highlights the most important decision-making stakeholders involved in the development and use of a new product or product feature.

The design team should note key decision makers in a tabular format, before proceeding to the next step.

Step 3a – Cycle of use analysis

The purpose of this step is to assess different stakeholders' value drivers and incentives with respect to behavioral, technical, and social attributes. This step of the analysis should be performed from the usability perspective, with functional requirements grouped into "use cases" or procedural steps that users will perform while using a product or feature. The output of this analysis is a list of stakeholders, ranked according to the degree to which each is affected by a particular attribute, from a behavioral/usability, technical/utility, and social perspective.

Behavioral/Usability attributes focus on the degree of behavior change a user would have to make in order to use or adopt a specific product or new feature, in comparison to existing practices or procedures. A score of 0 is assigned for no change, one for minimum change, and five for a completely new way of accomplishing a specific task.

Technical/Utility attributes center on a product's technical functionality and performance. The ranking for each feature-stakeholder combination is based on how critical the stakeholder considers a particular feature to be. Another way to view this is with respect to the value a stakeholder places on the technical benefit (utility) of a new feature. A "must have" feature is scored 5, and a "nice to have" feature is scored 1, using a tabular assessment.

Social Interests are attributes that consider social factors, such as power and prestige, associated with using a particular product, technology, or product feature. This ranking is based on the stakeholder's perception of status associated with using a specific product or feature. In scoring social interests, a perceived large increase in social status is assigned a score of 5, and minimum or no change is assigned a score of 1. The process of combining scores in each area is captured in detail by Agrawal et al. (2012).

Step 3b – Monetary flow analysis

This stage of the analysis assesses stakeholders with financial interests in a product in order to identify key decision makers who would influence product adoption from a financial perspective. In this step, the team starts with a list of stakeholders from step 2 (CVCA), and documents constituents that pay for products or new product features. This involves quantifying the financial gain or loss that each stakeholder would potentially experience, given the introduction or adoption of a new product or feature. If a stakeholder could potentially experience a large financial loss (relative to other stakeholders), then he/she would be assigned a score of −5. On the other hand, if a stakeholder could potentially gain financially from the introduction of a new product or feature, then he/she would be assigned +5. If a stakeholder is not affected financially then the score is 0. At the end of this exercise, the team will have a financially focused stakeholder-product features table with scores ranging from −5 to +5 for each product.

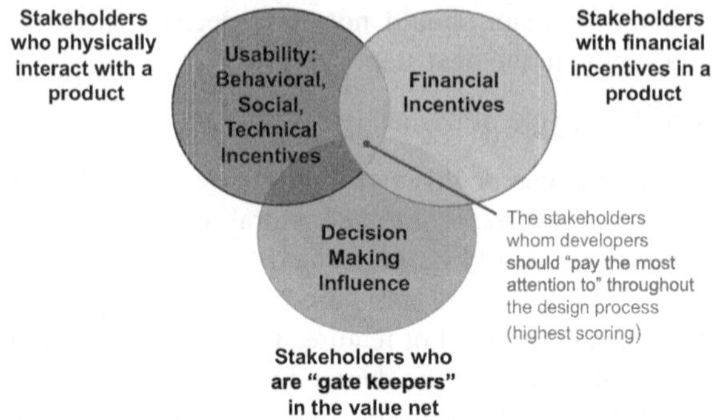

Fig. 1 Schematic representation of stakeholder attributes

Step 4 – Final stakeholder ranking

The final analysis step facilitates designers in determining which stakeholders to involve in the product design and development process, and provides recommended user interaction strategies (per the design methods discussed in Sect. 2.1).

At this stage, the design team has three ranked tables – one from the decision-making perspective, the product use perspective, and the financial perspective. From this information, designers can differentiate between stakeholders with an interest in feature development (from a usability perspective), versus those who may be influential in driving product adoption. Stakeholders with over-lapping interests at the intersection of usability incentives, financial incentives, and decision-making influence, are those whose needs designers should dedicate the greatest amount of time and resources to (as schematically highlighted in Fig. 1).

With the stakeholder ranking information, a design team will be able to segment stakeholders into the following four categories:

2.3.1 High Ranking: In Decision-Making, Influence Usability, and Financial Incentives

As previously noted, it is essential to include stakeholders from this category in the design and development process, since they have financial and decision making power that drives product adoption, and are likewise involved in the physical use of a new product or product feature. It is recommended that stakeholders in this category work alongside design team members through participatory and iterative design methods (Schön-type), as well as through scientific methods such as focus groups, surveys, and the minimum viable product strategy (Simon-type interactions).

2.3.2 High Ranking: In Usability

Stakeholders in this category will be closely involved in actual product usage – either as end users, intermediate operators, or maintenance personnel, etc. These stakeholders will typically be individuals or groups who physically interact with a product and hence should be actively involved in the iterative, hands-on design and development process. To enhance product usability and functionality, stakeholders in this group should ideally engage in product design through methods such as participatory or inclusive design (as discussed in Sect. 3.1).

2.3.3 High Ranking: In Decision-Making Influence and/or Financial Incentives

Stakeholders in this category have a high degree of decision-making power and a vested financial interest in a new product or product feature. But these stakeholders are not involved in the direct physical use of a product (i.e. they have 0 ranking in the usability matrix). Since these stakeholders have no direct interaction with products but do have control over finances and/or product purchasing, it is recommended that their inputs be considered using the scientific/technical (Simon-type) design methods.

2.3.4 Low Ranking in Two or More Categories

Stakeholders who receive low scores in two or more categories are considered lower priority stakeholders. In terms of resource allocation, we recommend that the design team avoid allocating a large percentage of resources toward the development of products or product features to satisfy the needs of stakeholders in this group (unless their needs correspond with the needs of stakeholders addressed previously).

Following this four-step ranking and stakeholder categorization process, the team can utilize the framework to include high priority stakeholders in the product development cycle, in an effort to satisfy the needs of users/stakeholders with the greatest degree of influence over product use and adoption.

3 Phase II: The Influence of User Expertise on Product Innovation

The second phase of research is an ongoing study that focuses on the following research questions:

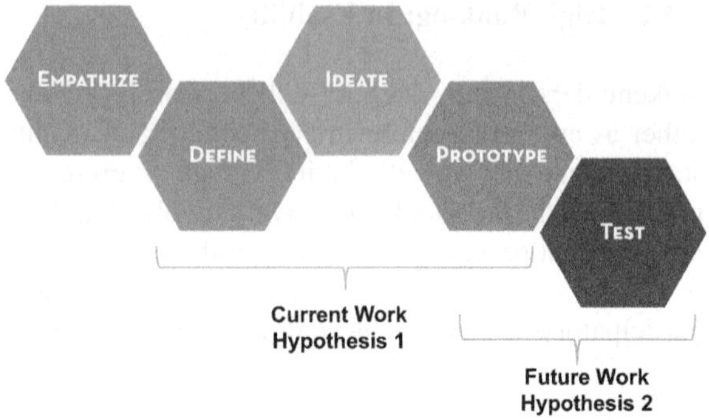

Fig. 2 Schematic representation of early stage design phases

Within each targeted user group, which subjects should one design with and for during the product development process – a technologically adept expert user or the "average" user?

Does designing with and for particular users and user groups have an impact on the usability, functionality, novelty, adaptability, and cost-benefit of new products and systems?

To examine these questions, we are studying the roles of expert and novice product users at discrete phases of the product development process, with an emphasis on early stage concept design and ideation (per Fig. 2). The goal of this research phase is to evaluate the degree to which designs differ, in terms of factors such as usability, functionality, and cost-benefit, when they are based on inputs from novice versus expert users. As previously noted, this research stems from an earlier study by Shluzas (2011), which illustrated that medical device developers often prefer to collaborate with industry thought leaders during the conceptual design and product testing phases of development. However, surgical procedures developed primarily based on input from lead-user surgeons (von Hippel 1986) frequently resulted in the development of technology that was difficult for a broad range of users to embrace.

In the current study, we hypothesize that at the conceptual phase of product design both expert and novice users will have a higher preference for design concepts based on input from expert users. This is due to the ability for experts to identify needs ahead of the target market (von Hippel 1986), and because experts have been shown to have a better understanding of functional relationships (Chi et al. 1981), and are better at identifying problem solving strategies (Klein 1999). However, at the usability and functional testing phases of design we hypothesize that both experts and novices will prefer designs that have been improved based on inputs from novice users. This is attributed to novice designs potentially being less reliant on user skill, more intuitive, and more adaptable to different usage scenarios.

To test these hypotheses, we chose to examine the design of intramuscular (IM) drug delivery devices, namely syringes, since these products involve daily interaction by users of varying skill levels. More specifically, these devices were selected because they are small on one hand, but also complex mechanical systems. Their usage involves multiple stakeholders with different design requirements (e.g. patients, caregivers, regulators, and manufacturers), and recruitment and access to expert and novice users (e.g. nurses with varying levels of training and experience) was feasible. Since syringes are designed for multiple use environments, studying their design provides an opportunity to examine the design of products intended for variable usage scenarios.

The experimental protocol, to examine the initial hypothesis involving user interaction for the design of early stage product concepts, involves six steps:

1. Research team recruits expert and novice users (n = 9–12 users per group).
2. User Input Sessions: Users provide needs and inputs, and rank their top 5 needs.
3. Research team separately aggregates and codes needs from the novice and expert user groups.
4. Designers (n=2) create computer aided design (CAD) models, based on design inputs from users of each group.
5. Users from each group select preferred design concepts.
6. Research team maps user preferences to designs based on inputs from either the expert or novice user groups.

3.1 User Input Sessions

For each user, we conducted a 60 minute user input session at either the Stanford Center for Design Research (CDR) or the user's place of employment. During these sessions, users were asked to complete an initial survey to document their injection experience and indicate the demographics of patients to whom they have administered injections. Users were then shown a brief video that illustrates the injection process, in order to frame the design exercise in the context of administering injections within a clinical setting. For about 30–45 minutes, users described their wants and needs through sketching, written, and oral descriptions. These sessions were videotaped to fully capture user inputs, and to provide inputs to designers at a later time. Finally, users completed the session with a post survey, in which they were asked to rank their top needs and document their technology adoption profiles.

3.2 Data Coding and Concept Generation

The user inputs were divided into novice and expert user groups. For each group, three overarching themes were synthesized and each user input (i.e. comment) was manually coded to gain a sense of the main design priorities by *quantity* of comments. To supplement each designer's empathy and understanding based on user inputs alone, designers researched IM injection details online (google) and watched online videos of IM injections (youtube). To generate designs, each designer selected quotes that "stuck out" as significant and representative of the users in each group.

For our cohort of novice users, three guiding priorities emerged from the coded data: (1) Safety, (2) Efficiency, and (3) Accuracy. Using these design priorities, designers first generated miniature concept hand-sketches. These mini-concepts represented small fragments of features (not fully functional designs). They next created two to four detailed composite sketches (fully functional syringes) on letter-sized paper that combined the most promising features. These detailed storyboard sketches served as the basis for generating 3D CAD renderings (in Rhino3d), corresponding technical drawings (in Solidworks), and an annotated design brochure (in Powerpoint). The process was repeated for expert users, but with the new priorities of: (1) Safety, (2) Accuracy, and (3) Efficiency.

Examples of two designs created based on inputs from the novice and expert user groups are shown in Figs. 3 and 4, respectively.

3.3 Mapping User Preference to Designs Based on Expert or Novice Inputs

Based on the top ranked design concepts from users of each group, we aim to map if novice users prefer designs based on inputs from novice users or expert users, and if expert users prefer designs based on inputs from either novices or experts, per the diagram in Fig. 5. Selection criteria include factors such as usability/perceived ease of use, functionality, speed and efficiency, adaptability, nurse safety, patient safety, and cost-benefit.

The intended theoretical and practical contributions of this research phase are: (1) to build on research in lead user innovation, by von Hippel (1976, 1986) and others, by illustrating the unique contributions that novice and expert users make at discrete phases of the design process, and how their contribution impacts product adoption; (2) to contribute to literature on design adaptability and intuitive design by illustrating which user groups facilitate the design of products for variable use scenarios; and (3) to extend literature in participatory design, from the human computer interaction and software development fields, to the design of tangible products. As a final goal, we aim for the designs created in this study to be used as future drug delivery devices in routine clinical care.

4 Discussion and Future Direction

Our research shows that there is substantial dissent in the nature and concept of 'users' among product development and design communities. Though often seen as individuals, users are frequently part of a complex stakeholder network that has been integrated into the development process through a series of interconnected relationships.

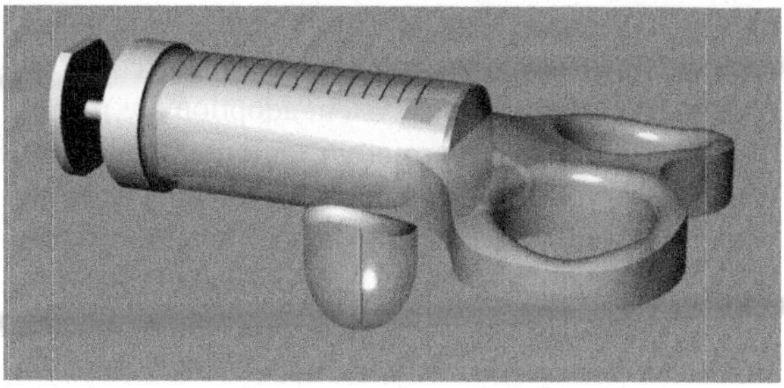

Fig. 3 Design concept (syringe) based on novice inputs

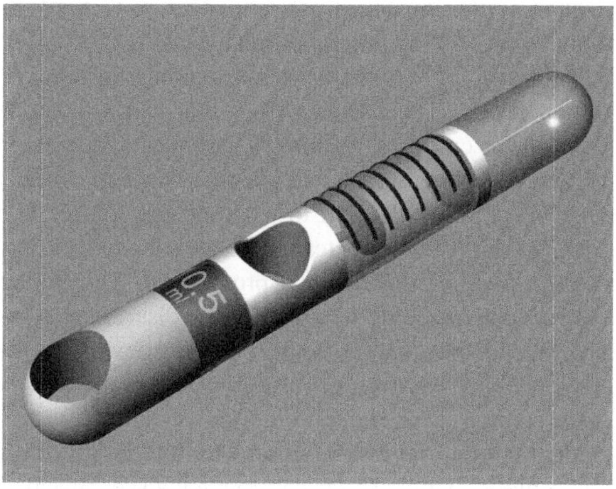

Fig. 4 Design concept (syringe) based on expert inputs

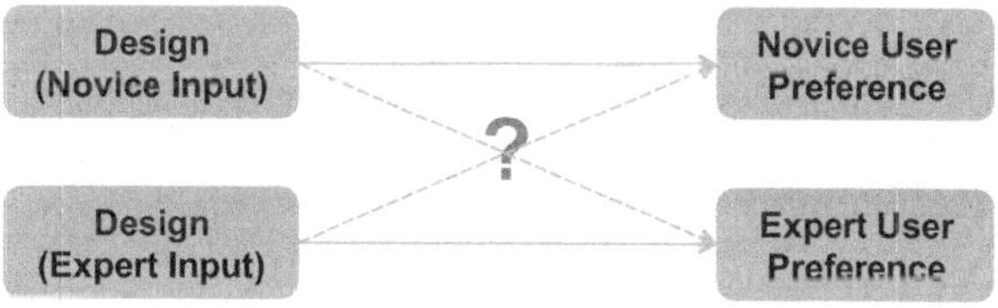

Fig. 5 Mapping user preference to designs based on inputs from novice or expert users

Phase I of our research demonstrates that in order to increase the success of products at a sub-system level (i.e. in terms of usability, functionality, social-status or financial gain), it is necessary for designers to identify all product stakeholders and target development efforts to satisfy the needs of constituents with the greatest degree of influence over product use and adoption. To accomplish this, we propose a four-step framework that provides a comprehensive system for evaluating stakeholder groups and allocating resources accordingly. Contrary to current processes for assessing and prioritizing users and stakeholders that are often fragmented and qualitative, the proposed *Design Stakeholder Identification, Assessment and Ranking Framework* combines several tools into one systematic process. It is geared toward the design and development of complex systems, involving an embedded network or ecosystem of users and stakeholders. This framework provides designers with a tool for ensuring that the needs of stakeholders with a high degree of influence over product/system adoption are met.

In Phase II we are currently examining the role of expert versus novice users in early stage product design. This work stems from earlier research (Shluzas 2011), which showed that medical device developers often prefer to collaborate with industry thought leaders in the conceptual design and product testing phases of development. However, surgical procedures developed primarily based on input from lead-user surgeons (von Hippel 1986) were shown to result in the development of technology that was often difficult for a broad range of users to embrace. By evaluating the impact of user expertise on technology development, this research aims to provide an improved understanding of user and stakeholder roles in the development of complex systems, and new insights regarding user-centric product design teams. The findings from this work should enable the research team to postulate an empirically founded hypothesis regarding which user groups to include in the conceptual design and ideation phase of product development.

In the future, we plan to examine the role of expert and novice users at the usability and functional testing phases of product development (i.e. for the development of three-dimensional prototypes). This future work intends to provide a clearer understanding of user roles at discrete phases of the product development process. In the context of user-centric design for the development of complex products and systems, we also aim to expand and test the insights gained from this research in different industry domains. We are particularly interested in evaluating the role of user groups for the design of products intended for adaptive and rapidly changing user environments (i.e. sustainability engineering and information technology).

References

Abras C, Maloney-Krichmar D, Preece J (2004) User-centered design. In: Bainbridge WS (ed) Encyclopedia of human-computer interaction. Berkshire Publishing Group, Great Barrington, pp 54–59

Agrawal V, Vaidya A, Shluzas LA, Steinert M, Katila R (2012) Conceptual and practical user integration into the design process – a four step stakeholder approach. In: Dorian M, Mario S, Neven P, Nenad B (eds) Proceedings of the 12th international design conference (Design 2012). Dubrovnik

Chi MT, Feltovich PJ, Glaser R (1981) Categorization and representation of physics problems by experts and novices. Cogn Sci 5(2):121–152. doi:10.1207/s15516709cog0502_2

Donaldson KM, Ishii K, Sheppard SD (2006) Customer value chain analysis. Res Eng Des 16:174–183

Klein G (1999) Sources of power: how people make decisions. MIT Press, Cambridge

Lüthje C (2003) Characteristics of innovating users in a consumer goods field: an empirical study of sport-related product consumers. Technovation 24(9):683–695

Moore GA (1991) Crossing the chasm: marketing and selling high-tech products to mainstream customers. HarperCollins Publishers, New York

Rogers EM (1962) Diffusion of innovations. Free Press, Glencoe
Rowley TJ (1997) Moving beyond dyadic ties: a network theory of stakeholder influences. Acad Manage Rev 22(4):887–910
Schön D (1983) The reflective practitioner: how professionals think in action. Basic Books, New York
Shah S (1999) Sources and patterns of innovation in a consumer products field innovations in sporting equipment. MIT Sloan School, Cambridge
Shluzas LA (2011) The influence of design and development practices on outcomes: a case-based analysis of medical device design. Stanford University, Dissertation
Shluzas LA, Leifer LJ (2012) The insight-value-perception (iVP) model for user-centered design. Technovation. doi: 10.1016/j.technovation.2012.08.002
Shluzas LA, Steinert M, Leifer LJ (2011) Designing to maximize value for multiple stakeholders: a challenge to med-tech innovation. In: Cully SJ, Hicks BJ, McAloone TC, Howard TJ, Dong A (eds) Proceedings of the 18th international conference on engineering design (ICED'11). Technical University of Denmark (DTU), Copenhagen, 15–18 Aug 2011
Simon HA (1976) Administrative behavior: a study of decision-making processes in administrative organization, 3rd edn. The Free Press, London
Suh NP (2001) Axiomatic design: advances and applications. Oxford University Press, New York
Ulwick AW (2002) Turn customer input into innovation. Harv Bus Rev 80(1):91–97
von Hippel E (1976) The dominant role of users in the scientific instrument innovation process. Res Policy 5(3):212–239
von Hippel E (1986) Lead users: a source of novel product concepts. Manage Sci 32(7):791–806
Urban G, von Hippel E (1988) Lead user analyses for the development of new industrial products. Manage Sci 34(5):569–582

Part IV
Make Ideas Tangible

Connecting Designing and Engineering Activities

Thomas Beyhl, Gregor Berg, and Holger Giese

Abstract Different design thinking activities result in a multitude of analog as well as digital artifacts. These capture the working results and are employed as a medium to communicate and preserve the embodied design decisions, observations and insights. When engineers or design thinkers want to revisit particular design activities, the information captured by the latest artifacts typically handed over or maintained are not enough. In addition, earlier artifacts, their context, dependencies between artifacts, the design rationale and other related details would be required. However, this information is often hard or impossible to recover if it was not systematically captured and documented. In the first year of our research project we therefore studied how to organize the design artifacts and their dependencies in a cost-effective manner to be able to retrieve information for engineers who have to realize the results. This includes an understanding of the actual challenge concerning documenting during design thinking.

1 Introduction

In today's business, different stakeholders interact to bring new and innovative products to market. Figure 1 depicts an overview of the typically involved stakeholders: the client, the design thinking team, the engineering team, the manufacturer and the end users. Typically, the client's management issues a design challenge to a consulting company (1). Within this consulting company, the design thinking team goes through the different stages of the design thinking process (Plattner et al. 2009) and interacts (2) accordingly with the prospective end user of their product or service idea. We focus on product ideas here. After iterating this

T. Beyhl • G. Berg • H. Giese (✉)
System Analysis and Modeling Group, Hasso Plattner Institute for IT Systems Engineering at the University of Potsdam, Helmert-Street 2-3, D-14482 Potsdam, Germany
e-mail: holger.giese@hpi.uni-potsdam.de

H. Plattner et al. (eds.), *Design Thinking Research*, Understanding Innovation, 153
DOI 10.1007/978-3-319-01303-9_11, © Springer International Publishing Switzerland 2014

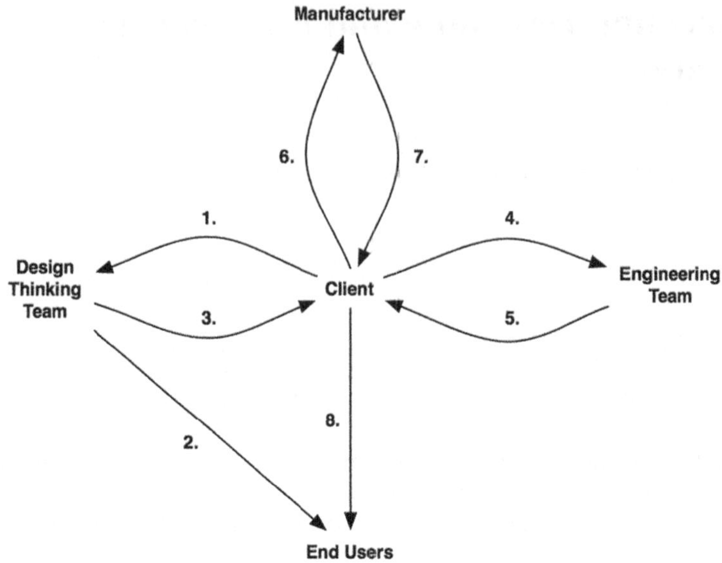

Fig. 1 Illustration of the collaborations between different stakeholders involved in design thinking projects

idea several times, the design thinking team hands over this idea to the client's management (3). How this handover looks depends on the client's wishes and may range from an informal presentation to a formal specification based on a template provided beforehand. Afterwards, the client's management decides about the realization of the product. If the client's management favors the realization of the product, the client uses in-house engineers or outsources it (4). Depending on the setting, the client combines engineering and manufacturing. Therefore, the client's management passes the available information about the product idea to the engineering team. Further, the engineering team realizes the desired product as possible within engineering constraints and hands back a blueprint to the client (5). If the client's management is pleased with the blueprint, it is passed on to manufacturing (6 and 7). Then, the product can be brought to market (8). The steps concerning the blueprint (5) to bringing the product to market (8) are in general uncritical concerning communication due to standardization, e.g. standard notations like UML.[1] Within our research project we investigate the steps concerning the impact of the handover (3) on the subsequent realization of the product idea (4). While practitioners have established best practices, these steps are still critical due to non-standardization. If the outcome of these steps is insufficient, it is likely that the product being manufactured is a less desirable product, which does not fulfill end user needs, in contrast to products being manufactured by engineers who are well informed.

[1] http://www.uml.org

To summarize, innovative ideas cannot unfold their full potential if their realization does not conform to elicited requirements due to uninformed decisions made by engineers. Engineering is the last step in the overall process before the actual manufacturing takes place and the new innovative product is brought to market. The involved engineers are responsible for realizing these desirable products, which are to be feasible as well. Therefore, engineers need to be empowered to make well-informed decisions concerning realization alternatives and tradeoffs between desirability and feasibility. But typically design thinkers only present the final ideas to the engineering team or merely to the client's management – their journey of how the design thinkers arrived at their final ideas is often neglected. Often, such a presentation mainly conveys features instead of the underlying requirements or design constraints. Besides making a final presentation, design thinkers also often present a final design prototype that covers important aspects regarding desirability. However, feasibility aspects can only be covered partially by design thinking team members with a background in the involved engineering disciplines.

A design prototype can be considered a model that fulfills a certain purpose (Gabrysiak et al. 2010). Thus, a design prototype also abstracts from properties that are irrelevant to conveying the final idea. These abstractions might lead to misinterpretations during engineering, which is usually focused on feasibility, since its result is always feasible. Still, trade-offs during engineering might lead to a less viable or even less desirable product. Figure 2 depicts that a successful idea needs to be desirable for the end users, feasible to realize and also economically viable.

From an engineering perspective, the dimensions of desirability and feasibility are the most important. If desirability or feasibility are not fulfilled, then also viability is not given – neither can you sell an undesirable product nor can you manufacture an unfeasible product viably. Figure 3 depicts a possible tradeoff space, which consists of the three dimensions mentioned by (Brown 2009). The intersection of all three sets depicts all desirable, feasible and viable product alternatives. A design prototype has a desirable design, but covers engineering constraints only partially. Thus, only presenting and handing over a design prototype to engineers is insufficient. But at least, it is a good starting point, since it provides an overview of the idea. Still, engineers require a trace of which alternatives have been explored, which constraints have been uncovered and how these influence the subsequent decisions.

Edelman and Leifer (2012) described two distinct modes of *path determination*. Either a decision is made in-situ, based on the readily available information (*way finding*), or the design thinkers plan ahead by navigating through the design space "along the most efficient route" (*navigation*). Thus, the design thinkers' journey progresses differently depending on the available information. During this journey, different prototypes are built. As argued by Tohidi et al. (2006), creating and evaluating multiple prototypes in parallel yields better results compared to employing only a single one. Consequently, by restricting the documentation on the final prototype, much valuable information discovered from building and evaluating the other ones are only passed on implicitly and might have to be re-discovered later on by the engineers.

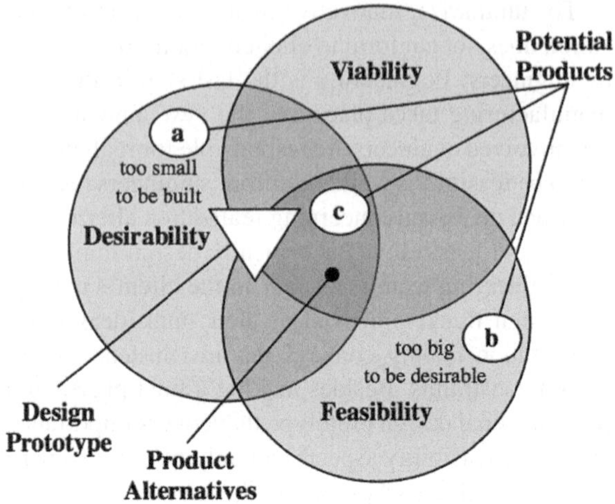

Fig. 2 Successful ideas cover all three dimensions adapted from (Brown 2009)

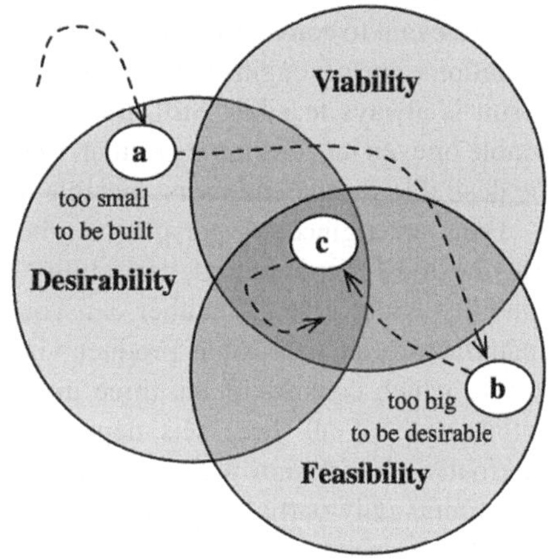

Fig. 3 A design thinking team's journey between desirability, feasibility and viability adapted from (Brown 2009)

Traceability can support design thinkers to capture their journey. Gotel et al. describe traceability for requirements as *"the ability to describe and follow the life of a requirement, in both a forwards and backwards direction"* (Gotel and Finkelstein 1994). Further, (Gotel and Morris 2011) specify the basic concepts of traceability: signs, tracks and traces. They define a sign as *"an identifying mark made by, or associated with for a particular purpose, an animate or inanimate object"*. They continue by defining a track as *"a pattern of signs created as these signs are generated"*. "To trace" means "to identify a track following its pattern sign by sign". These definitions can be applied to design thinking. Design thinkers have to make signs to create a track, which can be traced by engineers later on to recover the design thinkers' journey (Fig. 3) and, more importantly, which decisions they made. Traceability is easy if the available documentation is perfect. However, at least in students' projects, documentation is often neglected and incomplete (Bernd and Gluchow 2012), which makes traceability a difficult task.

In Sect. 2, we present the current state of the art. Then, Sect. 3 describes the current practice of documenting in educational and professional settings. Afterwards, Sect. 4 describes how the insights we gathered led us to reframe the challenge that we are investigating. Based on the reframed problem, we describe the experiments we conducted in Sect. 5. Section 6 describes our envisioned documentation framework. Then, Sect. 7 concludes the report and outlines future work.

2 State of the Art

In software engineering, *traceability* is employed to follow the lifecycle of development artifacts. Traceability helps to understand requirements by providing contextual information to manage their changes and to verify requirement satisfaction (Gotel and Morris 2009). Winkler et al. provide an overview about traceability in requirements engineering (Winkler and von Pilgrim 2010). In our approach, we focus on traceability approaches as used in requirements engineering, since these approaches are closely related to the needs of design thinking. Gotel and Finkelstein (1997) define traceability as the *"ability to describe and follow the life of requirements in both a forwards and backwards direction"*. In general, a clear distinction between *pre-requirements specification traceability* and *post-requirements specification traceability* can be made (Winkler and von Pilgrim 2010). The elicitation, discussion and validation of requirements, before they are included into the requirements specification document, are defined as pre-requirements specification traceability. The implementation of these requirements during the design and coding of a software system, on the other hand, is referred to as post-requirements specification traceability. Based on these definitions, design thinking has to deal with pre-requirements specification traceability and engineering deals with post-requirements specification traceability. Moreover, design thinkers and engineers manage traceability links (traces) of an informal nature concerning our overall challenge. Therefore, we focus on non-functional traces as defined by Pinheiro (2003).

In the related literature, multiple traceability approaches[2] for creating and maintaining traces are discussed. These approaches can be classified into runtime creation of traceability links, e.g. (Jouault 2005), recovery of traceability links, e.g. (Grechanik et al. 2006), (Marcus and Maletic 2003), (Egyed 2001), (2003) and

[2] Due to space limitations, we are not able to describe the large number of available traceability approaches in detail.

(Hayes et al. 2003), and combined approaches, e.g. (Poshyvanyk et al. 2006). Further, traceability approaches exist capturing the hierarchy and context of traceability links, e.g. (Seibel et al. 2010) and (2011).

Gotel and Morris (2009) state that agreed requirements are the result of information content transformations between different representations, e.g. use cases can be derived from the transcription of an interview recording. Unfortunately, such transformations are usually not bidirectional and important information can get lost. Therefore, Gotel et al. argue that relationships, i.e. traceability links, between artifacts might not be as obvious as one might expect concerning syntactic and semantic issues. They conclude that *"storing, using and maintaining extensive media-rich materials is far more costly than creating them in the first place."* This also applies to design thinking.

As the literature conveys, traceability is important – not only in software engineering (Gotel and Morris 2011). However, not everybody is willing to apply traceability techniques and maintain traceability links (Arkley and Riddle 2005). The same is true for capturing the rationale behind design decisions (Klein 1993).

3 Current Practices

First of all, we investigated different application scenarios combining design thinking and engineering. Further, we took a step back and looked at the current practice in educational and professional design thinking settings using surveys, interviews and explorative experiments.

3.1 Application Scenarios for Design Thinking

When speaking about connecting design and engineering activities, the overall process as depicted in Fig. 1 has to be considered as well. The following Figs. 4, 5 and 6 depict possibilities of how design thinking and engineering can be connected (Beyhl et al. 2012).

In the *decoupled setting* in Fig. 4 design thinking (step 1 to 3 in Fig. 1) and engineering (step 4 and 5 in Fig. 1) take place sequentially and a clear separation between design thinking and engineering exists. In this setting, design thinkers and engineers (or at least the management) meet rarely, mainly to present and hand over the results (e.g. design prototype, documentation of the final result). Since the final idea is stable it is known what kind of engineer is required. Therefore the idea has to be conveyed to the engineering team. As for all presented settings, the engineering team is usually larger than the design thinking team. Thus, a small set of design thinkers has to hand over their knowledge to the larger engineering team. Of course, the introduction of new personal is not effortless and requires documentation and communication (Balzert 1997). Design thinking teams are multidisciplinary and,

Fig. 4 Decoupled setting

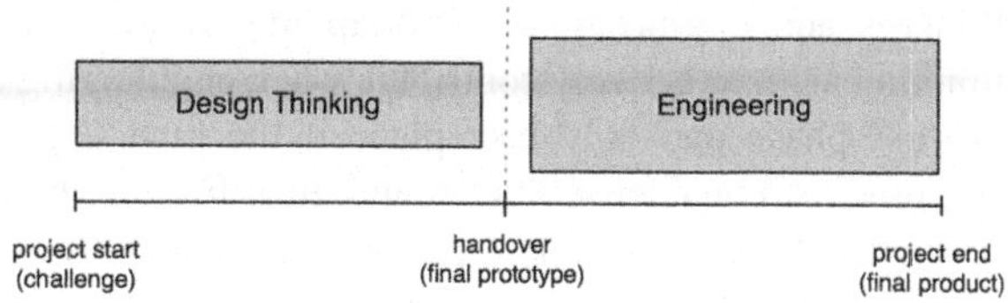

Fig. 5 Overlapping setting

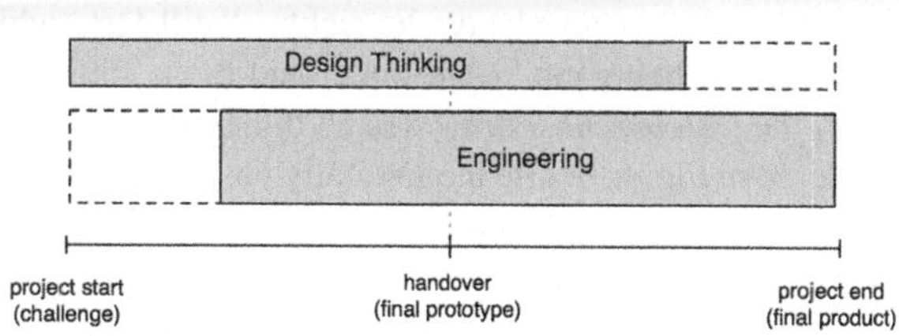

Fig. 6 Concurrent setting

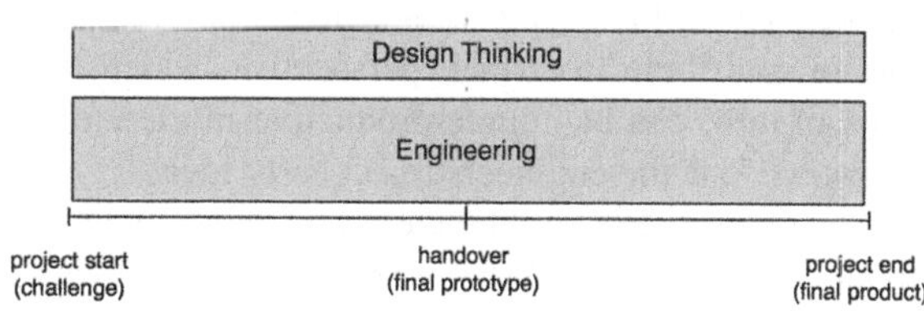

therefore, also contain engineers. In this decoupled setting, these engineers are still different from the engineers, who realize the idea later on. Thus, the engineers of the design thinking team cannot act as knowledge carriers. Still, they are able to provide feasibility assessments throughout the design thinking process. This decoupled setting is the common approach for the subsequent steps of design thinking and engineering. Therefore, the progress speed can be considered as normal and less agile concerning the product idea, which is stable after the handover takes place. In practice, we mainly observed the decoupled setting – especially since the final documentation in professional settings is considered to be sufficiently complete.

Figure 5 depicts the *overlapping setting*. This setting does not separate between design thinking and engineering. Instead, design thinkers and engineers work together before and after the handover of the final product idea. Consequently, the engineers get insights into the design thinkers' knowledge and the design thinkers are available for questions later on. As for the decoupled setting, the engineering team is also larger than the design thinking team. By introducing the overlap between design thinking and engineering, the overall team size increases, which requires more communication between the team members than before. Therefore, an additional process overhead is introduced which might slow down the overall process. Furthermore, the actual engineering should not start until the product idea is stable. Consequently, the overlapping setting has no benefits concerning progress speed. Besides the necessary communication between design thinkers and engineers, the overlapping setting allows engineers of the design thinking team to actively contribute when engineering takes place. The overlapping process phase can only take place if the kind (e.g. product or service) of final outcome is considered stable and therefore it is known which engineers are required. If the engineers start the implementation of the idea too early, these efforts might be wasted if design thinkers come up with a different idea later on. Or, as Alexander stated, simultaneous development and implementation of requirements cannot always be right – "still less, while concrete is being poured and metal is being cut" (Alexander and Beck 2007).

In the *concurrent setting,* design thinkers and engineers work simultaneously as depicted in Fig. 6. While theoretically possible, this setting is unrealistic concerning team size, progress speed and contribution. Design thinking teams are small and additional participating engineers would increase the team size. Also if only a few engineers participate, the knowledge they gained has to be shared with the rest of the engineering team. Thus, the handover problem is merely shifted. Furthermore, engineers from the engineering team are not trained in design thinking and are only able to contribute from their perspective, which is usually focused on feasibility. This, in turn, can be counterproductive in view of the brainstorming rule *"Defer judgment!"*, if the engineers reject early ideas.

Design thinkers, on the other hand, are only able to contribute their knowledge about the product idea during engineering. Moreover, if the engineers start too early with the realization of an immature idea, they have to adapt their result as soon as the design thinkers shift their direction, which introduces additional overhead. Further, it has to be known what kind of engineers are required in advance. However, design thinking projects are open-minded and the kind of final outcome is not known in advance.

3.2 *Educational Settings at the D-School Potsdam*

We investigated the current practice at the HPI School of Design Thinking (D-School) in Potsdam,[3] Germany to get an understanding of how students document their findings. The information manager of the D-School provides a set of tools and templates as best practices, which should be used by the students. The D-School provided a wiki as documentation platform during our first interviews. The students used this wiki to write down their findings, upload photos of analog artifacts, e.g. post-its or physical prototypes, and add comments to them. The D-School only prescribed the structure up to the project level. Each team was responsible for structuring their project site. They often used a timeline (e.g., week 1, week 2, etc.) or process structure (e.g. understand, observe, point of view, etc.) concerning the design thinking process (Plattner et al. 2009). Besides

[3] http://www.hpi.uni-potsdam.de/d_school/home.html?L=1

the wiki as documentation platform, the D-School provides a file system share. Analogue to the wiki, the information manager of the D-School only prescribes the structure up to the project level and the students are responsible for structuring their project folder as they like. They often use the same structure as with the wiki. The students use the file system share to upload and share big files such as audio or video recordings and presentation slides. Currently, the D-School is switching from the wiki to Incom[4] as the communication and documentation platform. Within Incom the students document their findings, upload photos and add comments using a built-in timeline feature. In general, our interview partners favor this timeline in contrast to the wiki. The main advantages of Incom are the visualization of data and feedback sharing. In comparison to the previous wiki, pictures are displayed, not only stored as files. Besides that, the built-in timeline feature helps the D-School faculty to comment on the project progress. Shortcomings of Incom for the system administrators are the time-consuming setup and the lack of support to maintain/adapt the platform to specific needs. Besides these software tools, the D-School also provides templates, e.g. presentation templates, with suggestions of what to write down. For example, the D-School provides a question template with questions to be answered every day. During our interviews, it turned out that documentation is created relatively close to the end of the project and before the final presentations take place. Nonetheless, the students do not only present the final outcome, i.e. their final prototype. Although they present up to six times their project progress, the documentation of the project's journey is often neglected. Depending on the project partner, the students describe their idea in more detail afterwards, e.g. further meetings with the project partner or final reports. We further investigated how the students rate the provided best practices. The following statements are an excerpt from our interviews:

- First, the problem is nobody is willing to document.
- Incom is not a blog for sure, it's too complicated, it's too official, it's no fun to work with it.
- There is no individualization in Incom.
- Incom is no collaboration instrument.
- I would like to have something like Google[Docs].
- A login screen is a barrier at the beginning.
- It's nice to see what other are doing [...], but there are too many subsections I have to open [...].
- After two weeks we start to forget where things came from.
- The communication should take place in Incom, but it's done via e-mail.
- The best would be to document during the process, but we don't have time for it.
- Some people document during the process and some at the end.
- At the end of a 12 week project we don't know what we did at the beginning.

To summarize, documenting alongside the project is disliked and students often see no benefit for themselves, although they acknowledge its importance. Further, the provided tools are often too difficult to use and do not enable collaboration. Instead, the students use a collection of different tools depending on the agreement

[4] http://www.incom.org

of the project partner. Among others, this collection consists of GoogleDocs[5] for collaborative editing of text documents and spreadsheets, Dropbox[6] for document sharing and easy photo upload from mobile devices and Facebook[7] to arrange meetings. Each of these choices indicates use cases and scenarios that are not yet covered otherwise. Thus, by combining these use cases and the information they affect into an overall tool, a documentation platform can be realized that students are willing to use. As experience shows (Gabrysiak et al. 2012b), providing students with a well-structured template of what to document is not sufficient as they will only look at this template after they finished collecting all the insights and establishing their ideas.

3.3 Professional Settings at D-LABS

Based on our examination into how the software design consultancy D-LABS[8] gathers and manages their insights (Gabrysiak et al. 2011), a bachelor project investigated how these findings and results are documented (Gabrysiak et al. 2012a). D-LABS is a start-up company located in Potsdam (Germany). They provide expertise in design thinking for complex multi-user software products to customers as a service (cf. 1, 2, and 3 in Fig. 1). Their projects are run based on a catalogue of design thinking methods adapted to the domain of software.

Figure 7 depicts the Design Led Innovation (DLI) approach (Mecklenburg 2012). D-LABS covers the requirements engineering stage of a project. To do so, they start by understanding the problem domain through a 360° view. Afterwards, they start end user research using observations and interviews. The succeeding synthesis phase during which the point of view is created, iterated and refined is the most important one, since the ideation of solution concepts relies on the correctness of these agreed-upon findings. Based on the ideation, the prototyping stage usually starts with paper prototypes, which are iterated till a prototype of the graphical user interface is created.

Concerning the application scenarios for design thinking, D-LABS fits into the decoupled setting (Fig. 4). Their engineers are responsible for the creation of interactive prototypes, which enables them to assess the corresponding feasibility. However, these engineers are not part of the engineering team later on. In order to maximize the reuse capabilities of their results, apart from a detailed documentation of how they got there, D-LABS also provides re-usable software engineering artifacts from which the final product may be generated. As in all projects, D-LABS uses a shared folder for all documents, such as pictures, persona

[5] http://docs.google.com

[6] http://www.dropbox.com

[7] http://www.facebook.com

[8] http://www.d-labs.com/english/

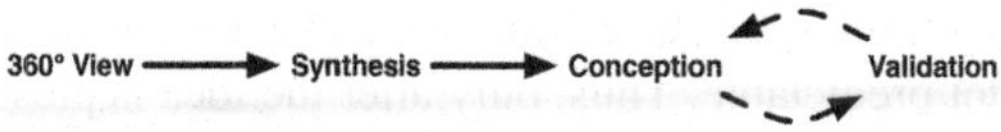

Fig. 7 D-LABS' Design led innovation (DLI) approach (http://www.d-labs.com/english/angebot/prozess/)

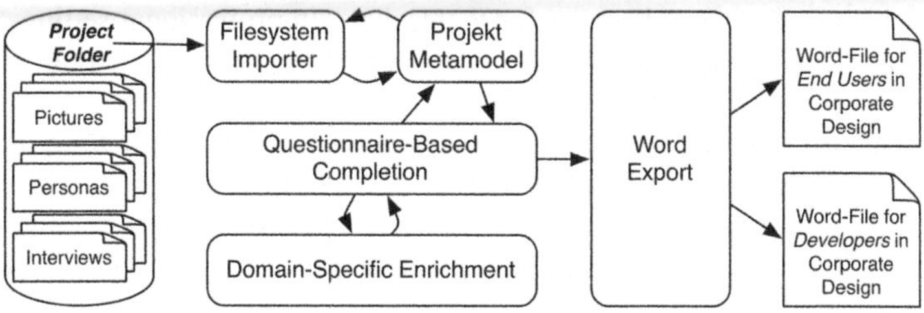

Fig. 8 The resulting tool of the bachelor project, capable of producing specifications suitable for multiple distinct perspectives (Gabrysiak et al. 2012a)

descriptions and digital prototypes. These folders are structured the same. Further, file versions are managed manually.

In the bachelor project "Design Thinking meets Requirements Engineering" the students deduced a meta-model capable of describing the overall state of a project by systematically going through the corresponding project folder. Thus, with such an approach, as illustrated in Fig. 8, standards for documentation can be defined and enforced. Of course, enforcing a sophisticated documentation style introduces overhead and might even decrease project performance. However, since the documentation being handed over is what the customer pays for, it is essential that the overall idea is captured, documented, and represented thoroughly.

In a study conducted as part of this bachelor project, software engineering students were asked to evaluate design thinking deliverables. They had to focus on whether they would be able to implement the specified idea. One of the results indicated that while it is important for design thinkers to have a process that they can rely upon when explaining how they arrived at their results (Lyon 2011), most of the software engineering students did not look at the attached description of the design thinking process. Instead, they argued, that they trusted results presented by consultants as long as their process is specified (Knolle 2011).

3.4 Summary

Current practices show that the decoupled setting dominates since design thinking and engineering are two steps that can clearly be separated. Furthermore, it is rarely feasible that design thinkers accompany the whole engineering process later on. When design thinking projects are initiated and a design challenge is issued, the result is usually not predictable. Thus, only after the idea is presented, is it known which engineering experts from which domain are required to successfully implement the idea. This temporal dependency restricts the application of the overlapping or concurrent setting.

We also have to distinguish between the educational and the professional setting. In educational settings, the design thinkers (students) perceive the documentation effort as high and without benefits for themselves. In professional settings, on the other hand, design thinking professionals assess the documentation effort as medium but necessary, since they rely on their documentation to create business value for their customer. The successful application of design thinking in professional settings is usually found in combination with structured templates and artifacts. By customizing the documentation to the needs of engineers, the realization of such ideas can be simplified to increase the chances of a successful realization.

4 Reframing the Problem

As the current practices indicate, a framework[9] for documenting design thinking projects is needed. Such a documentation framework is the pre-condition for providing traceability information for engineers. Figure 9 depicts the *initial* black box view of our research challenge. Based on input from design thinkers, we initially hypothesized that we would be able to increase the overall documentation quality for engineers. Since the input and output were unspecified at that point in time, we started to investigate the current practices of documentation alongside the project (input) and the documentation passed on (output). To get an understanding of what kind of information is available and how traceability can be applied, we also looked at the needs of design thinkers and engineers. Therefore, the documentation framework might consist of methodologies and rituals as well as adaptable and combinable software tools.

4.1 *Applying Traceability in Design Thinking*

When design thinkers synthesize information from different sources, they also re-represented it in other forms and dimensions. Consequently, the originating sources are hard if not impossible to recover if no traceability is used (Gotel and Morris 2009). Initially, we argued that traceability techniques may also be

[9] We do not understand this documentation framework only as a framework from engineers' perspective, but also as a framework from design thinkers' perspective.

Fig. 9 Black box view of the envisioned documentation framework

applicable for design thinking projects to hand over the traceability information along with the documentation to the engineers later on.

Hypothesis H1

Applying traceability to design thinking helps engineers to create a desired product within engineering constraints.

Figure 10 depicts how the knowledge of design thinkers and the knowledge of engineers can be linked with the help of traceability. The knowledge base of engineers increases if traceability is applied due to the ability to trace back and get the rationale behind design thinkers' decisions (b). If traceability is not applied, the knowledge of engineers remains on the level of the basic idea (a).

Applying traceability helps to follow diverging and converging activities of design thinkers and to understand the rationale behind their decisions. This is done in a similar way as in traditional engineering by tracing back and forward. As defined by Winkler and Pilgrim (Winkler and von Pilgrim 2010), "backward traceability refers to the ability to follow the traceability links from a specific artifact back to its sources from which it has been derived." Thus, engineers may reconstruct the context in which the artifact was created to understand the rationale behind undocumented decisions. Forward traceability, on the other hand, stands for "following the traceability links to the artifacts that have been derived from the artifact under consideration." In design thinking, we only consider non-functional traces, because of its informal nature. Pinheiro classifies non-functional traces in four groups: reason, context, decision and technical (Pinheiro 2003). Based on this classification, we investigate what to capture during design thinking projects. Moreover, software is required to support traceability, otherwise maintaining traceability information would be a manual and error-prone task (Winkler and von Pilgrim 2010). In such a case, traceability would be worthless. Depending whether traceability has to be applied in educational or professional settings, different conditions have to be fulfilled. For educational settings, the amount of *explicit knowledge* that is written down is much smaller than the amount of *implicit knowledge*, as illustrated in Fig. 10.

Explicit knowledge is the knowledge that is written down by design thinkers. Implicit knowledge is the knowledge that is not written down. Implicit knowledge only exists in the design thinkers' heads and is only manifested in prototypes. In professional settings, on the other hand, the explicit knowledge is already documented in a way that works best for the company and their clients. Still, new

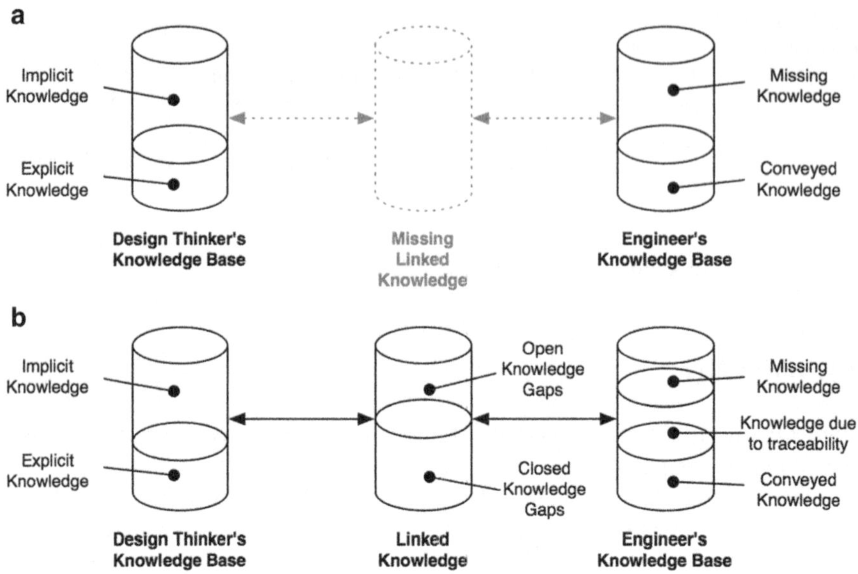

Fig. 10 Link the knowledge of design thinkers and engineers

concepts are difficult to evaluate in professional settings since financial risks have to be avoided and corporate secrets must remain secret. Therefore, we decided to focus on educational settings first and transfer successful concepts to professional settings later on. Consequently, applying traceability is only the second challenge, especially in educational settings.

4.2 Getting Design Thinkers to Document in Educational Settings

The first challenge of our research topic is to get design thinkers to document their findings, insights, alternatives and decisions. They almost exclusively rely on analog artifacts such as post-its and tangible prototypes, which are captured in digital artifacts, for example photos and videos. They upload only the most important artifacts to the used documentation platform, especially in educational settings. In professional settings, common file system structures lead to a more complete and structured set of artifacts.

Usually, the person who uploaded the artifact does not comment these digital artifacts further since digital artifacts already carry different (meta) information. However, the same is assumed for the analog information within the digital artifact. Depending on the type of digital artifacts, this can be a problem later on due to technical constraints, for example handwriting recognition is a difficult task.

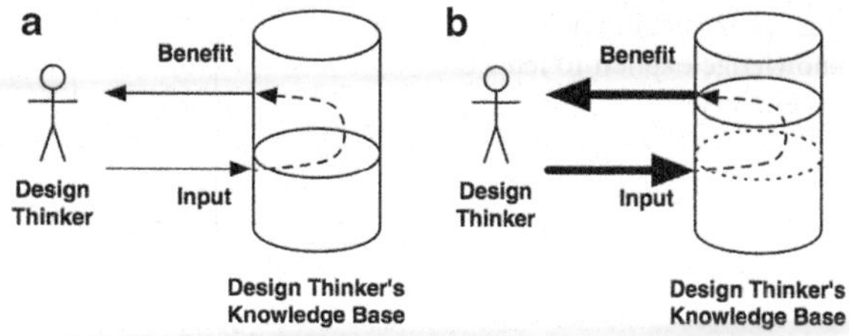

Fig. 11 Provide benefits for design thinkers to get them to document

Moreover, knowledge carriers, for example design thinkers, are not willing to document. The reason is that someone who documents thoroughly is usually not one of the beneficiaries of the documentation – especially in educational design thinking settings (Fig. 11a). In the literature, different benefit problems are described, for example the *traceability benefit problem* (Arkley and Riddle 2005) and the *rationale capture benefit problem* (Klein 1993). Instead, they document the way they are accustomed to which is based on their academic background. Further, different kinds of information need to be captured and traced, e.g. process information, artifact information and context information. By making documenting a beneficial activity, design thinkers would intrinsically be motivated to document appropriately (Fig. 11b). The question is how such benefits should look.

Hypothesis H2

If the students' benefits are high, the amount of documentation is high.

4.3 Structuring the Informal World of Design Thinkers

Based on our interview results, design thinkers write down their knowledge in natural and visual language, e.g. sketches, when it is conscious. In general, while being the only specification language "that can be assumed to be common to all the stakeholders", natural language is also inherently ambiguous (Gervasi and Zowghi 2005). In contrast, engineers prefer a common standardized structure and formal precision without a margin of misinterpretation.

Hypothesis H3

Chronologically documented knowledge is less helpful for engineers than knowledge following formal documentation standards.

Therefore, the question comes up how to present and access the knowledge from design thinkers' and engineers' points of view. Due to the informal world of design thinking, the resulting ambiguity has to be taken into account. "Uncertainty is [also]

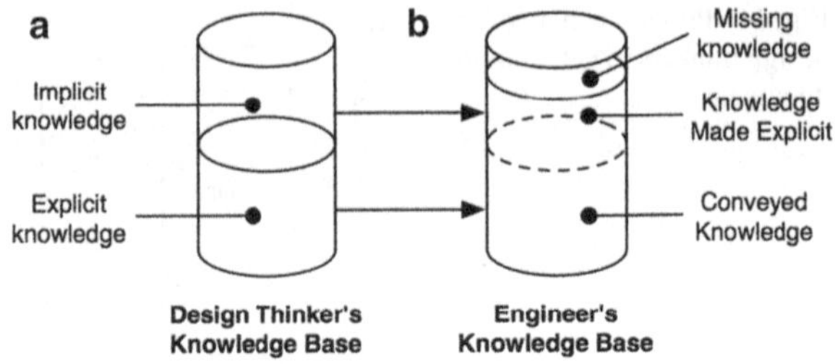

Fig. 12 Make implicit knowledge explicit to reduce uncertainty

inherent and inevitable in software development processes and products," as stated by the uncertainty principle in software engineering (Ziv et al. 1997). We hypothesize that appropriate combinations of traceability approaches can help to reduce such uncertainty by making implicit knowledge (Fig. 12a) explicit and therefore enhance the knowledge base of engineers (Fig. 12b).

Hypothesis H4

Traceability reduces uncertainty by making implicit knowledge explicit.

4.4 Summary

To summarize the challenges and characteristics described above, Fig. 13 depicts how design thinkers and engineers interact with the envisioned documentation framework. Design thinkers document their insights mainly using digital artifacts such as photos, videos and text documents while engineers request necessary information like requirements, use cases and traceability links. Before such a system can be built, problems of documenting have to be tackled, such as it being perceived as boring, slowing down the process and without personal benefit.. Especially design thinkers do not profit from their own documentation, since they only rarely refer to it later. Those who profit are the engineers not the design thinkers themselves. Consequently, design thinkers are purely extrinsically motivated to document their insights. Even worse, being instructed to document may disturb a moment's momentum.

The first challenge is to just get design thinkers to document at all by providing them with their own benefits, especially in educational settings. When sufficient documentation exists the second challenge of applying traceability can be tackled. As third challenge, the captured traceability information can be used to make implicit knowledge explicit.

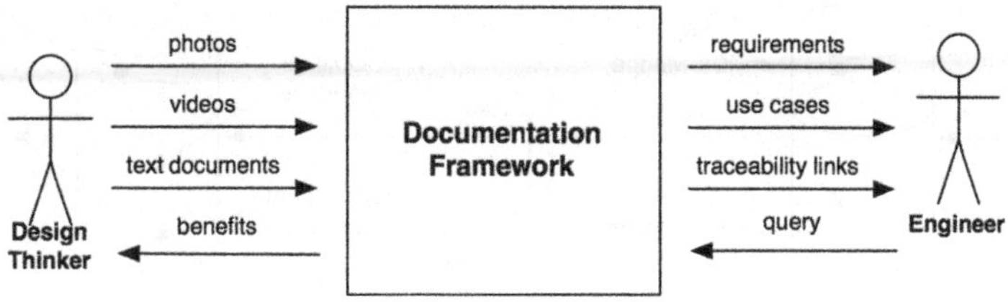

Fig. 13 Black box of the documentation framework

5 Experiment Results

We investigated the underlying hypothesis H3 to provide evidence for the existence of the handover problem by conducting two experiments. While one experiment explores the appropriateness of different kinds of handover documents, the other one investigates how design thinkers and engineers structure and organize their artifacts. We describe these experiments in more detail in the following sections. Based on the evidence for hypothesis H3, the other hypotheses will be evaluated in the forthcoming year of the project.

5.1 Which Handover Document Is the Most Appropriate One?

We set up an experiment to compare handover documents of different kinds to accept or reject the hypothesis H3 (Beyhl et al. 2012). Figure 14 depicts how handover documents pass on explicit knowledge of design thinkers to engineers. Depending on the kind of the handover document, the passed on information might be incomplete.

Therefore, we prepared three equivalent handover documents: (a) a video of a design thinkers' presentation (15 min), (b) an informal document (5 pages) and (c) a 15-page software requirements specification (SRS). Each of these handover documents describes the same final idea about a documentation robot capable of gathering information within design thinking sessions as well as a web interface to access the information collected by the robot. The informal document and the SRS were derived from the captured presentation and are considered equivalent. The captured presentation shows students presenting the final idea and prototype they came up with. As used in practice, the informal document presents the same final idea by using mockups and personas. Further, it describes the *as-is* process and the *to-be* process. The software requirements specification describes the final idea by describing requirements and use cases in a standardized manner. The overall sizes of the documents were kept small enough so that they could be studied in less than 30 minutes. A total of 30 computer science students from different institutions

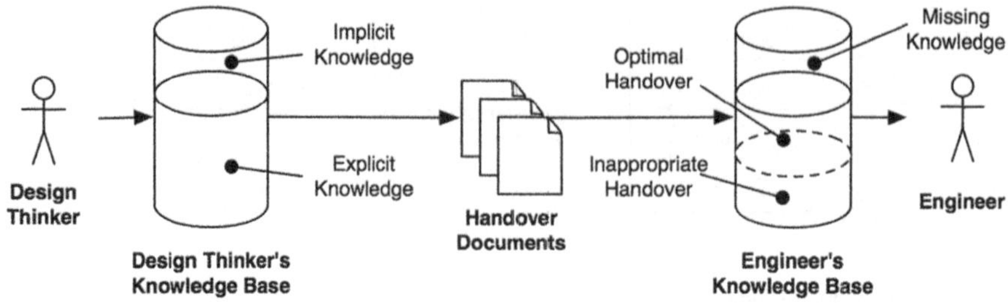

Fig. 14 Investigation as to which handover document most fits the concerned engineer's needs

participated. Each document was studied by ten of the participants. For each document, these participants were balanced concerning their experience (e.g. academic degree, practical experience) in computer science and their field of study (e.g. software engineering, informatics, business informatics). Each participant read the document in less than 30 minutes and answered a questionnaire afterwards. The participants were allowed to look things up in the handover document assigned to them while answering the questionnaire.

Going through the assigned document and answering the questionnaire took less than one hour for all of the participants. Figure 15 depicts the most important results. Among other questions, the participants were asked to rate their agreement to the following statements in a Likert scale (1 means very strong disagreement, 7 means very strong agreement).

Question A: *"I'm able to create a product that fulfills the requirements of the customer."* The participants rated the software requirements specification (average of $\overline{x} = 5.0$, variance of $s^2 = 2.7$) as the most suitable handover document concerning the ability to create a product that fulfills the requirements of the customer, because this kind of documentation fits best for software- and hardware system. We were surprised that the participants rated the video ($\overline{x} = 3.7$, $s^2 = 2.9$) higher than the informal document ($\overline{x} = 2.4$, $s^2 = 1.8$). A possible reason might be that such a presentation video conveys an illusion of sophistication and customer contact. Usually, such an implicit assumption is incorrect.

Question B: *"In the document, all requirements are covered sufficiently."* For this question, the participants rated the SRS as most sufficient concerning the specified requirements ($\overline{x} = 3.9$, $s^2 = 4.3$), again. Although the SRS is rated best, the Likert value is still relatively low. As initially stated, we had to restrict the documents' number of pages to ensure that the participants were able to read the document within an acceptable time. In software projects, such documents can have hundreds of pages, depending on the complexity of the product. In such a case, we believe that engineers would rate the SRS even higher. The captured presentation ($\overline{x} = 2.2$, $s^2 = 2.0$) and the informal document ($\overline{x} = 2.4$, $s^2 = 4.7$) are considered to have an even lower coverage of the requirements. From an engineer's point of view,

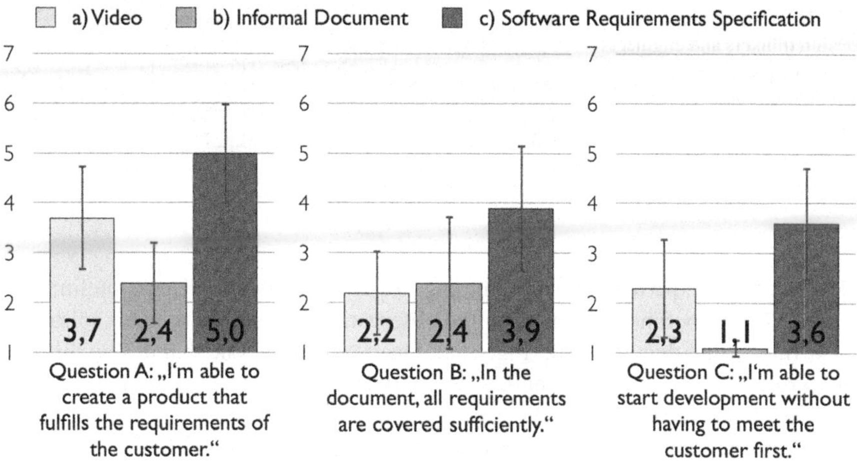

Fig. 15 Selected results (average values) of our experiment concerning the appropriateness of handover documents (1 means very strong disagreement, 7 means very strong agreement)

this is not surprising, since these two kinds of documents are only expected to convey the basic idea.

Question C: ***"I'm able to start development without having to meet the customer first."*** Also for this statement, the SRS received the highest rating ($\overline{x} = 3.6$, $s^2 = 3.4$). The statement implies a sufficient completeness of the handover document and, therefore, correlates with the other two statements. Again, the captured presentation ($\overline{x} = 2.3$, $s^2 = 2.7$) is rated higher than the informal document ($\overline{x} = 1.1$, $s^2 = 0.1$). As argued before, as opposed to the informal document, the video might convey customer contact, which makes further meetings superfluous.

To summarize, we consider these results as a trend that showed captured presentations and informal documents to be insufficient as handover documents. This is because of their informal and non-standardized nature, which does not cover all necessary insights and ideas. Nonetheless, the captured presentation and informal documents are a good starting point to pass on the basic idea and to provide a common understanding of the idea.

5.2 A Preliminary Qualitative Comparison of How Design Thinkers and Software Engineers Organize File Structures

Besides the investigation of the appropriateness of handover documents, we investigated how design thinkers and engineers organize their digital artifacts and, thus, file structures by conducting a second experiment concerning hypothesis

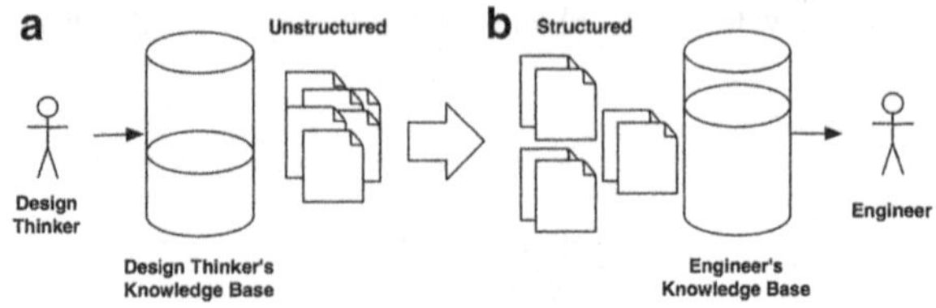

Fig. 16 Comparison of how design thinkers and engineers organize their knowledge base

H3. Since this experiment is still ongoing, we are only able to discuss preliminary results. This experiment is based on the observation that design thinkers follow no clear rules while documenting (Fig. 16a). One reason might be that design thinkers often rotate the task of documenting. In contrast, engineers have common rules on how to document (Fig. 16b). We are conducting this experiment to explore how a common structure for both, design thinkers and engineers, might look.

We prepared a set of digital design thinking artifacts consisting of whiteboard photos containing post-its, personas, interview transcripts, photos of prototypes and audio/video recordings. Overall, we provided 160 digital artifacts placed into one folder, which the students were asked to organize by using file operations (move, delete, rename, link, sort, etc.) provided by the file system explorer of the Windows operating system. The participants were assigned to groups of three. Each group consisted of one experienced design thinker, one person familiar with the design thinking method, who had not yet used it himself, and one person who had never heard of it before.

In the first part of the experiment, each participant had to individually organize the provided set of artifacts concerning the structure that she deemed most suitable. Afterwards, the participants answered a questionnaire about their chosen file structure, e.g. *"How do you rate the understandability of your file structure?"*, using a seven-point Likert scale.

In the second part of the experiment, the participants presented the file structures they created to each other. Afterwards, they were asked to discuss how a common file structure should look and created it using the provided files. Afterwards, they answered a subset of the questionnaire a second time, e.g. *"How do you rate the understandability of the new file structure?"* and *"Do you think that the understandability of the common file structure was increased due to the collaboration of you and your colleagues?"*, both using a Likert scale. Overall, the experiment had a duration of two hours for each group. In the following, we summarize preliminary observations and findings:

- D-School students recognized that the provided artifacts can be assigned to the design thinking method (Plattner et al. 2009). Thus, they created folders for each process step.
- Computer science students recognized parts of the software developments process and used a suitable file system structure.
- If they were not able to assign the artifact to a distinct process step, all students moved the corresponding artifacts into a folder named *"other"* or *"archive"*.

- During the discussion part of the experiment, the D-School student briefly explained the design thinking method to the computer science students using the created file structure. In general, the computer science students agree that this structure is the most convenient.
- In general, different digital artifacts often capture the same analog artifact, e.g. post-its. The students often discussed how to deal with such redundant information. Either they only kept one single final overview image and deleted all redundant pictures in-between or they kept all of them to be able to recover the evolution of artifacts later on in more detail.
- As opposed to the design thinking students, the software engineering students paid attention to details such as case sensitivity, underscores and whitespaces.
- Voting results were often encoded within file and folder names.
- Ranking of files and folders was enforced by employing prefixes such as numbers to encode for example the chronological order of design thinking method steps.
- The students rarely deleted, renamed, or duplicated any of the files.
- The practices we observed most frequently were:
  - Moving files into folders and only duplicating them if they can be assigned to different folders or, alternatively, creating links to single files instead.
  - Sorting files based on their type (e.g., image, text, or presentation) and recovering relationships between artifacts, afterwards.

These preliminary results indicate that design thinkers and engineers organize artifacts in conformance to the process they are accustomed to – either by following the design thinking method or the software development process of their choice. Further, the results indicate that engineers agree to the file structure proposed by design thinkers as the most suitable after having been introduced to the design thinking method. Nonetheless, both procedures should be supported to bridge the informal world of design thinking and the formal world of engineering.

5.3 Summary

Summarizing, the informal world of design thinkers has to be translated appropriately into the formal world of engineers. While design thinkers organize their artifacts following *implicit rules*, engineers have *explicit rules*, e.g. standards, how to document. Therefore, the envisioned documentation platform has to provide the ability to structure design thinking artifacts automatically concerning stakeholders' needs (Harzmann 2011; Tietz 2011). Parnas et al. state that it is sufficient to fake the documentation *"by producing the documents that we would have produced if we had done things the ideal way"* (Parnas and Clements 1986). Further, they state *"documentation that is written when the project is nearing completion is written by people who have lived the system for so long that they take the major decisions for granted.*" The authors conclude that "*the result is a document useful to people who know the system well, but impenetrable for newcomers*". Therefore, documentation should be created along the design thinkers' *idealized* journey from the authors' point of view.

6 Approach

Gotel et al. introduce the concept of sign makers (Gotel and Morris 2011). Applying this concept to our documentation challenge, design thinkers should act as sign makers. Later on, engineers can then rely on the created signs to trace decisions back to their underlying rationale. In design thinking, signs about *artifacts*, *method steps* and *contexts* can be made. Depending on design thinkers' profession, also *technical* signs can be created. Further, these signs have to be made permanent otherwise they are worthless.

6.1 Architecture

We already stated that the documentation challenge is manifold. To address each piece of the puzzle, a combination of solutions is necessary. Therefore, we suggest a configurable documentation framework suitable for design thinkers (e.g. rituals) as well as engineers (e.g. pluggable software tools) as depicted in Fig. 17. Four main pieces of the puzzle exist: *presentation framework*, *sensor framework*, *reasoning framework* and *active repository*. Each of these pieces can be divided even further; for instance, the sensor framework consists of different types of sensor, while the reasoning framework consists of different reasoning approaches.

From the design thinkers' perspective, the documentation framework has to be intuitive, easy to use, and highly customizable. Documenting has to be done unobtrusively without negative influences on the design thinking method. Further, the documentation framework has to provide benefits to design thinkers.

From the engineers' perspective, the documentation has to be complete and every decision of the design thinkers must be recoverable by engineers. Therefore, design thinkers' and engineers' knowledge has to be linked to a shared knowledge base. In traditional software engineering, developers use traceability approaches to link related content. The documentation framework has to apply such traceability approaches automatically to reduce the manual effort, because nobody creates traceability links manually (Arkley and Riddle 2005). Further, combining artifact, process, context and technical signs using such traceability approaches might enhance the quality of established traceability links by declining or confirming such links.

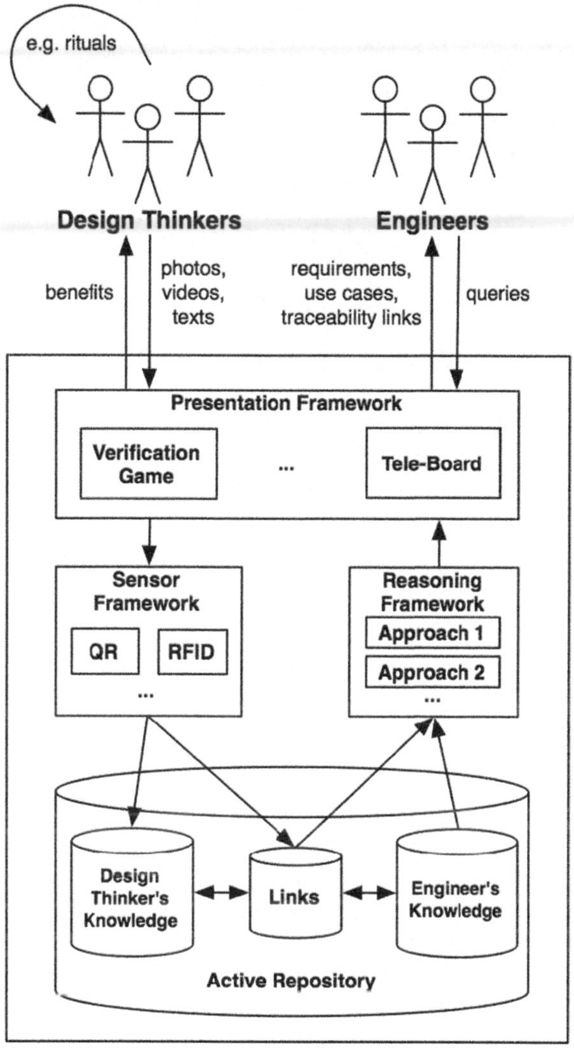

Fig. 17 Envisioned documentation framework

6.2 Presentation Framework

The presentation framework has to be easily adaptable since design thinkers are usually among the early adopters of new tools and services. Therefore, different software tools should be pluggable, e.g. Dropbox and GoogleDocs. Besides software, also new hardware tools should be easily pluggable, e.g. Tele-Board (Gumienny et al. 2012) to make use of electronic sticky notes.

To encourage design thinkers to document and make this task more exciting, games should be provided which makes documenting a less boring task. For

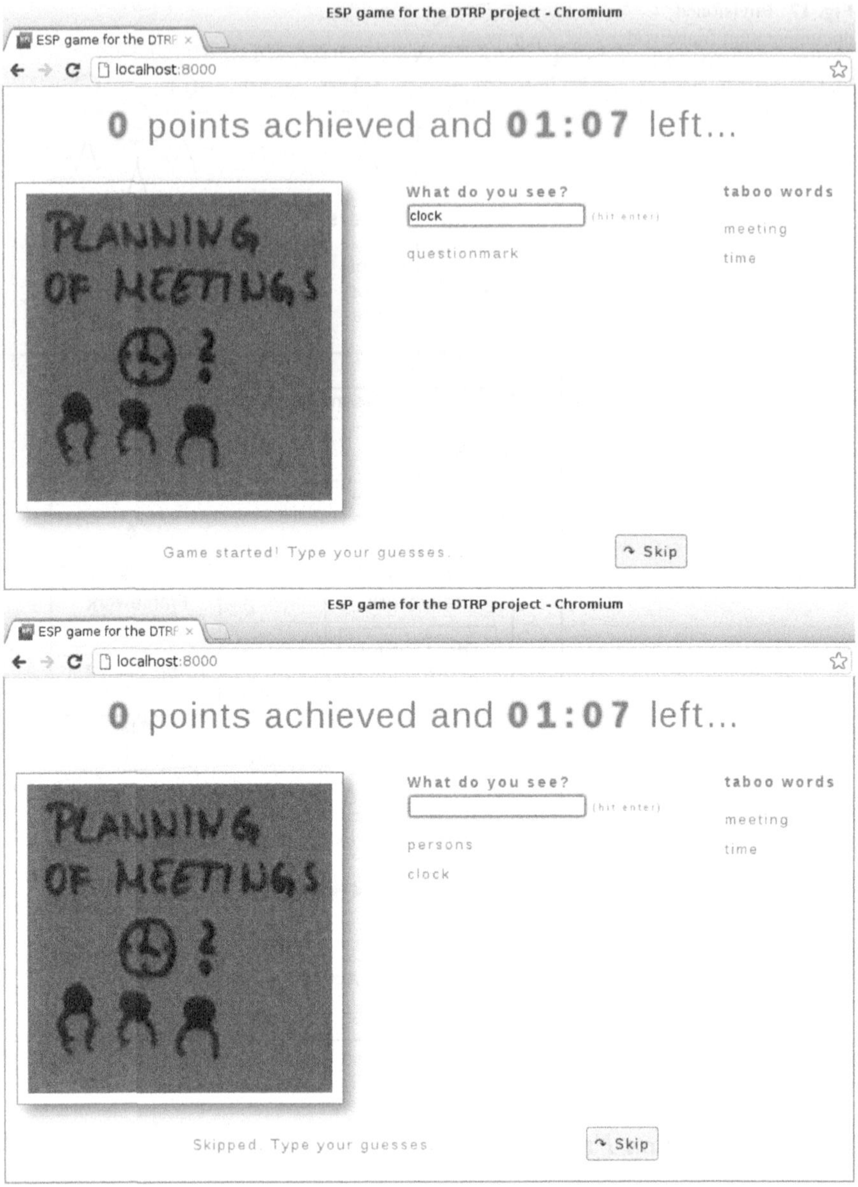

Fig. 18 Playful tagging of sticky notes

example, we currently implement a verification game (von Ahn and Dabbish 2004) for tagging images such as photos of post-its collaboratively. Within this game, independent players assign tags to the presented image until a first match is found. Due to this collaborative task, a common agreement is established and the presented images get a semantic based on the assigned tags. Later on, these tags can be used by design thinkers and engineers to look up information. This game makes non-searchable artifacts searchable. Figure 18 depicts a running game. The player on top has already guessed the word "clock", whereas the player below is guessing the same word. Besides making documenting a less boring task, this verification game exploits humans' capabilities to interpret images and avoids difficult and error-prone handwriting recognition. Moreover, design thinkers communicate their ideas visually. Subsequently, post-its only contain few texts and handwriting recognition would be inconsequential.

6.3 Sensor Framework

To capture information unobtrusively, sensors are used to track design thinkers' activities, the context of their activities and the artifacts that are created during their activities. Signs can be captured and made permanent by employing different sensors. Thus, besides the information that design thinkers actively upload onto the documentation platform, additional information is captured to make implicit knowledge explicit. Later on, engineers can use these signs to trace their questions to the collected information. In general, we distinguish between sensors for *process steps, artifacts,* and for capturing *contexts.* While *process sensors* capture information about the current process step and the activities within this step, *artifact sensors* capture artifacts such as personas, findings and alternatives. *Context sensors*, on the other hand, observe the overall context in which artifacts are created and decisions are made. Combining the different sensor information makes implicit knowledge explicit and complements the overall documentation.

For example, signs can be meta-information in images (e.g. timestamps or GPS locations) or (relative) positions of post-its on whiteboards. We already prototyped a detection approach for re-establishing post-it movements between whiteboards and post-it clusters. Figure 19 depicts exemplary post-it clusters. The QR[10] code at the bottom right side of each post-it enables our tool to detect the position of the post-it within the captured image. Based on this position information, clustering algorithms (Ertel 2009), e.g. *k-means* or *EM algorithm*, can be used to assign each individual post-it to a cluster. Each cluster represents a relationship between post-its, e.g. alternatives to realize the same end users' need – relationship information, which might be important for engineers later on. Additionally, our tool is able to extract each single post-it. Such post-its can then be used as input for the verification game. Further, post-its that occur multiple times in different contexts are considered as more important than other post-its.

Another sensor device, which may be integrated, is the Tele-Board (Gumienny et al. 2012), since it also supports the clustering of post-its and provides a history to

[10] Quick Response

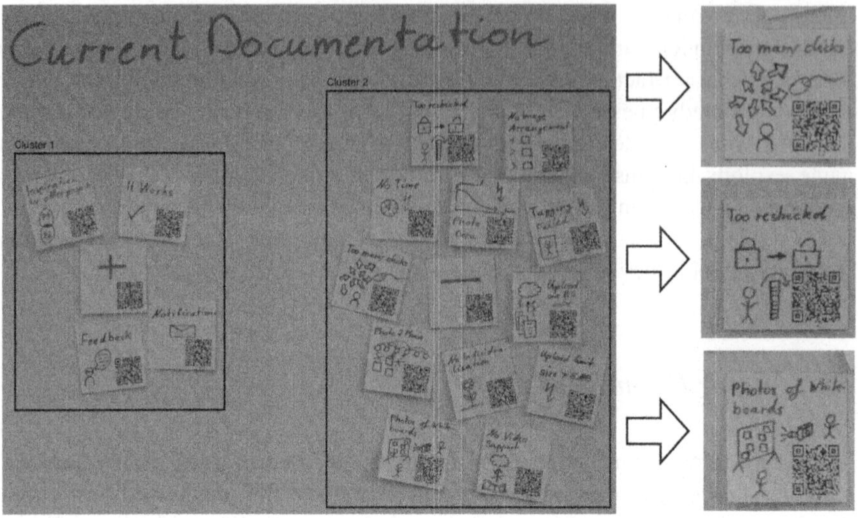

Fig. 19 Automatic post-it extraction and cluster detection based on QR codes

keep track of changes. Therefore, it already addresses traceability needs. In the forthcoming year, we are planning on prototyping additional sensors, such as position tracking of design thinkers within their workspace. Using RFID[11] technology, we might then be able to draw conclusions about their current activity. For example, when design thinkers stand nearby in front of a whiteboard, it is likely that they brainstorm, discuss or vote on their ideas.

6.4 Active Repository

After enough signs have been collected to re-establish the design thinkers' track, traceability techniques can be applied to follow the lifecycle of artifacts and enhance the overall quality of the documentation. To make the gathered information useful for engineers, an active repository combines the collected signs and enriches the gathered information in the background. Therefore, it performs analyses regarding the creation, validation/maintenance and deletion of traceability links. Already established traceability links might be input for additional analyses concerning the combination of traceability links. These analyses are performed in the background and the results are stored in the repository and can then be queried by engineers and design thinkers. For example, if the repository already contains the information that the design thinkers brainstormed or voted at a certain point in time (e.g. via position tracking) and, additionally, the design thinkers uploaded a

[11] Radio-Frequency Identification

photo of post-its with a similar timestamp and new QR codes, leads to a re-interpretation of the first information. In this case, it is safe to assume that the design thinkers brainstormed instead of voting. Due to the second information, the uncertainty remaining concerning the first information (brainstorming or voting) can be reduced. Depending on the kind of a traceability link, it may be necessary to maintain the link immediately or it can be sufficient to maintain the link later on concerning the *criticality* of the link. We regard this criticality as influence of the traceability link to query results. Further, the maintenance of one traceability link might involve additional maintenance of related traceability links due to their combination. In design thinking settings, only non-functional traces, e.g. traces about reason, context, process and technical issues, are considered.

6.5 Reasoning Framework

With the help of the reasoning framework, engineers are able to trace an artifact back to its creation and to follow its evolution. This tracing enables engineers to get the rationale behind design decisions within the underlying context. Therefore, they are able to recover the reasons for design decisions and make well-informed decisions during engineering. With such additional information at hand, engineers have a better chance of creating the desired product within engineering constraints.

7 Conclusion and Future Work

During the first year of our research project, we realized that the challenge is more manifold than we thought at the beginning of the project. Therefore, we reframed the problem and broadened the scope of our research. It turned out, that the first challenge is to get design thinkers to document, especially in educational settings, to evaluate our concepts concerning traceability in design thinking projects. The second challenge is to apply traceability techniques and combine their results to reduce uncertainties in the documentation. By doing so, the third challenge of enabling engineers to create the desired products within engineering constraints can be tackled.

In the forthcoming year, we will work on providing evidence for our hypotheses concerning the benefits of traceability in design thinking and engineering activities. Currently, our bachelor project *"From Creative Ideas to Well-Founded Engineering"* elicits requirements for the educational context and works on a first prototype of the envisioned documentation framework with a focus on the presentation framework. We are going to transfer approved concepts to professional settings later on.

Acknowledgement The authors are grateful for the input of Alexander Renneberg (D-LABS GmbH) and our student assistants Josephine Harzmann, Christoph Kühnl, Manuel Hegner and Lukas Pirl. The authors are also grateful for the input of Claudia Nicolai (D-School Potsdam), Harald Gögl (D-School Potsdam) and our bachelor project team *"From Creative Ideas to Well-Founded Engineering"*. The authors also thank Raja Gumienny for her support in preparing our experiments.

References

Alexander I, Beck K (2007) Point/counterpoint. IEEE Softw 24:62–65

Arkley P, Riddle S (2005) Overcoming the traceability benefit problem. Audio, transactions of the IRE professional group on, pp 385–389

Balzert H (1997) Software-management, Software-Qualitätssicherung, Unternehmensmodellierung. Lehrbuch der Software-Technik. Spektrum Akademischer Verlag GmbH, Heidelberg/Berlin

Beyhl T, Berg G, Giese H (2012) Tackling the documentation benefit problem in design thinking. In: Confestival 2012 - design thinking the future, Potsdam, pp 1–2

Brown T (2009) Change by design. HarperCollins, New York

Brügge B, Gluchow M (2012) Towards production ready software in project courses with real clients. In: Proceedings of the 1st international workshop on software engineering education based on real-word experiences (EduRex), Zurich, pp 5–8

Edelman JA, Leifer L (2012) Qualitative methods and metrics for assessing wayfinding and navigation in engineering design. In: Plattner H, Meinel C, Leifer L (eds) Design thinking research – measuring performance in context, Understanding innovation. Springer, Berlin/Heidelberg, pp 151–181

Egyed A (2001) A scenario-driven approach to traceability. In: Proceedings of the international conference on software engineering, Toronto, Canada, pp 123–132

Egyed A (2003) A scenario-driven approach to trace dependency analysis. Softw Eng IEEE Trans 29(2):116–132

Ertel W (2009) Grundkurs Künstliche Intelligenz: Eine praxisorientierte Einführung. Computational intelligence, 2nd edn. Vieweg + Teubner Verlag, Wiesbaden

Gabrysiak G, Edelman JA, Giese H, Seibel A (2010) How tangible can virtual prototypes be? In: Proceedings of the 8th design thinking research symposium, Sydney, pp 163–174

Gabrysiak G, Giese H, Seibel A (2011) Towards next generation design thinking: scenario-based prototyping for designing complex software systems with multiple users. In: Plattner H, Meinel C, Leifer L (eds) Design thinking: understand – improve – apply, Understanding innovation. Springer, Berlin/Heidelberg, pp 219–236

Gabrysiak G, Giese H, Beyhl T (2012a) Virtual multi-user software prototypes III. In: Plattner H, Meinel C, Leifer L (eds) Design thinking research – measuring performance in context, Understanding innovation. Springer, Berlin/Heidelberg, pp 263–284

Gabrysiak G, Guentert M, Hebig R, Giese H (2012b) Teaching requirements engineering with authentic stakeholders: towards a scalable course setting. In: Proceedings of the 1st international workshop on software engineering education based on real-word experiences, Zurich, pp 1–4

Gervasi V, Zowghi D (2005) Reasoning about inconsistencies in natural language requirements. ACM Trans Softw Eng Methodol 14(3):277–330

Gotel OCZ, Finkelstein CW (1994) An analysis of the requirements traceability problem. In: Proceedings of the 1st international conference on requirements engineering, Colorado Springs, CO, USA

Gotel O, Finkelstein A (1997) Extended requirements traceability: results of an industrial case study. In: Proceedings of the 3rd IEEE international symposium on requirements engineering, Annapolis, MD, USA, pp 169–178

Gotel O, Morris S (2009) More than just "Lost in translation". Software 26(2):7–9

Gotel OCZ, Morris SJ (2011) Out of the labyrinth: leveraging other disciplines for requirements traceability. In: 19th IEEE international conference on requirements engineering (RE), pp 121–130

Grechanik M, McKinley KS, Perry DE (2006) Recovering use-case-diagram- to-source-code traceability links

Gumienny R, Gericke L, Wenzel M, Meinel C (2012) Tele-board in use: applying a digital whiteboard system in different sit- uations and setups. In: Plattner H, Meinel C, Leifer L (eds) Design thinking research - measuring performance in context. Springer, Heidelberg/ New York/Dordrecht/London, pp 109–125

Harzmann J (2011) Questionnaire-based Completion of Models. Bachelor's thesis, Hasso Plattner Institute, University of Potsdam

Hayes JH, Dekhtyar A, Osborne J (2003) Improving requirements tracing via information retrieval. In: Proceedings IEEE international requirements engineering conference, Monterey, CA, USA, pp 138–147

Jouault F (2005) Loosely coupled traceability for ATL. In: Proceedings of the European conference on model driven architecture (ECMDA) workshop on traceability, Nürnberg, Germany

Klein M (1993) Capturing design rationale in concurrent engineering teams. IEEE Comput J 26 (1):39–47

Knolle L (2011) Vervollständigung von Anforderungen in Software-Design-Projekten. Bachelor's thesis, Hasso Plattner Institute, University of Potsdam

Lyon JB (2011) Balancing the Emic and the Etic: an ethnographer of design reflects on design ethnography. Innovation, Summer, pp 26–30

Marcus A, Maletic JI (2003) Recovering documentation-to-source-code traceability links using latent semantic indexing. In: Proceedings 25th international conference on software engineering, IEEE Computer Society, Portland, USA, pp 125–135

Mecklenburg C (2012) Eine Studie zur Nutzerzentrierung bei Innovation und Softwareentwurf. Master's thesis, Hasso Plattner Institute, University of Potsdam. In cooperation with D-LABS GmbH

Parnas DL, Clements PC (1986) A rational design process: how and why to fake it. Softw Eng IEEE Trans 2:251–257

Francisco Pinheiro (2003) Requirements Traceability. In: Sampaio do Prado Leite, Julio Cesar, Jorge Horacio (eds) Perspectives on software requirements. Springer, Berlin/Heidelberg, pp 93–113

Plattner H, Meinel C, Weinberg U (2009) Design thinking. Number ISBN-13: 978–3868800135. mi-Wirtschaftsbuch. In German

Poshyvanyk D, Gueheneuc YG, Marcus A, Antoniol G, Rajlich V (2006) Combining probabilistic ranking and latent semantic indexing for feature identification. In: 14th IEEE international conference on program comprehension, ICPC 2006, pp 137–148

Seibel A, Hebig R, Neumann S, Giese H (2011) A dedicated language for context composition and execution of true black-box model transformations. In: 4th international conference on software language engineering (SLE 2011), Braga

Seibel A, Neumann S, Giese H (2010) Dynamic hierarchical mega models: comprehensive traceability and its efficient maintenance. Softw Syst Model 9(4):493–528

Tietz D (2011) View-based transformations of requirements into different representations. Bachelor's thesis, Hasso Plattner Institute, University of Potsdam

Tohidi M, Buxton W, Baecker R, Sellen A (2006) Getting the right design and the design right: testing many is better than one. In: CHI '06: Proceedings of the SIGCHI conference on human factors in computing systems, ACM, New York, pp 1243–1252

von Ahn L, Dabbish L (2004) Labeling images with a computer game. In: CHI'04: Proceedings of the SIGCHI conference on human factors in computing systems. ACM Press, Vienna, Austria
Winkler S, von Pilgrim J (2010) A survey of traceability in requirements engineering and model-driven development. Softw Syst Model 9(4):529–565
Ziv H, Richardson D, Klosch R (1997) The uncertainty principle in software engineering. In: 19th international conference on software engineering (ICSE'97)

A Research Plan for the Integration of Design Thinking with Large Scale Software Development Projects

Thomas Kowark, Franziska Häger, Ralf Gehrer, and Jens Krüger

Abstract Design Thinking and agile software development processes are both widely adopted by innovative companies that try to create products with maximum end user value. Usually, the adopting companies work in smaller teams that tackle projects of manageable size, but huge companies are increasingly trying to adopt these methods in their large-scale projects as well. The main impediments for the adoption of the aforementioned methods in such settings are the strict requirements which, for example, enterprise software vendors have to fulfill. The need for complying with a plethora of mandatory product standards and providing comprehensive documentation of the development processes is not integrated naturally into the methodologies and, hence, tend to weaken their innovative potential. This chapter outlines our research agenda for a process model that combines agile software development processes, such as Scrum or Extreme Programming, with Design Thinking activities, while trying to maintain compliance with the previously mentioned product and process standards. Our initial process model is based on a series of expert interviews with people that previously applied Design Thinking in large companies, as well as on a thorough review of related work about similar approaches. We will further outline our evaluation process. This centers on a project based university course that transfers the idea of Stanford's ME310 to a computer-science-only setting. Investigating the teams' virtual collaboration activities, but also using traditional research methods, such as questionnaires and interviews, helps us to continuously improve our process model and prepare it for future test-runs within large partner companies.

T. Kowark • F. Häger • R. Gehrer • J. Krüger (✉)
Hasso Plattner Institute, University of Potsdam, 14482 Potsdam, Germany
e-mail: thomas.kowark@hpi.uni-potsdam.de; franziska.hager@hpi.uni-potsdam.de; ralf.gehrer@hpi.uni-potsdam.de; jens.kruger@hpi.uni-potsdam.de

H. Plattner et al. (eds.), *Design Thinking Research*, Understanding Innovation,
DOI 10.1007/978-3-319-01303-9_12, © Springer International Publishing Switzerland 2014

1 Introduction

Over the past years, Design Thinking has evolved into a powerful methodology to initiate and implement successful innovation. It especially excels in generating ideas and solutions that are beyond the incremental scope used by many companies. To achieve such radical innovation, Design Thinking is radical in its demands on collaboration, interdisciplinary teams, user centrism, and continuous alternation between divergent and convergent thinking. Following the example of the Stanford d.school,[1] similar schools were initiated worldwide,[2,3,4,5] to teach Design Thinking to students and professionals. Design agencies like IDEO not only use Design Thinking to reinvent or innovate their customer products, they also brought Design Thinking into companies like SAP (Savvas 2012), Microsoft (Lund 2011) or Apple (Thomke and Feinberg 2010). Furthermore, various articles in business and management journals (BusinessWeek 2004) (Beckman and Barry 2007) show that today Design Thinking is generally accepted as an innovation method.

Today, software engineering, especially for global enterprise applications, is becoming increasingly complex. Customers demand features faster and ideally want them tailored to their needs. Standards for business software, e.g., for security and accessibility, lead to a large amount of non-functional requirements, while laws and regulations from around the world need to be considered and integrated. Especially larger software companies introduce additional constraints on their teams and products to ensure high quality software and high availability of the software and its provided services. In this restricted setting, software companies are still expected to deliver innovative applications. In the past decade the software industry has evolved its engineering processes from strict waterfall process frameworks to more flexible agile or lean process models including methodologies such as Scrum (Schwaber and Beedle 2001) or Kanban (Anderson 2010). This change allows a faster reaction to changing requirements but does not automatically foster innovation. As Design Thinking has proven to be an effective methodology to create innovative ideas, the software industry is starting to include it into its processes. However, there has been little research in this area so far and the existing research is mainly comprised of interviews with Design Thinking workshop participants about their experience and single project case studies.

In this chapter, we introduce our research efforts to uncover useful ways of integrating Design Thinking into software development processes and to identify at which points in development cycles Design Thinking can excel and which tools and methods will enable the development teams to create innovative software that

[1] d.school Stanford – http://dschool.stanford.edu/.

[2] Aalto Design Factory – http://www.aaltodesignfactory.fi/.

[3] Aalto Tongji Design Factory – http://sfc.tongji.edu.cn.

[4] HPI School of Design Thinking Potsdam – http://www.hpi.uni-potsdam.de/d_school/home.html?

[5] d.school Paris-Est d.school – http://www.facebook.com/d.schoolparis

fulfills all necessary constraints and standards. The rest of this chapter is structured as follows:

To get an idea where the integration of Design Thinking in software companies currently stands, the next section provides an overview of existing integration approaches for Design Thinking and software development from both research and industry. Based on this overview, we explain our own research efforts and ideas in the third section of this chapter. The fourth section provides an overview of a Design Thinking computer science course that we are creating in order to evaluate our ideas and findings. How this evaluation will take place is described in the fifth section. The chapter concludes with a summary.

2 Overview of Existing Integration Approaches

As mentioned before, research to investigate how software development can benefit from Design Thinking is still scarce. This section provides an overview of existing research efforts as well as known approaches from companies. First investigations about the application of Design Thinking within IT Development have been made by Lindberg et al. (2012). They extracted 4 approaches to integrate Design Thinking within IT development activities:

- In the split project approach, Design Thinking is delivered as a service by a specialized Design Thinking team whose input can be used to start off the IT-development process.
- In the overlapping team model, one or two members of the development team should be integrated into the Design Thinking activities that take place before the actual software development to bridge the gap between the teams and prevent information loss.
- In the unified project model, the same team is working on Design Thinking and IT-development tasks. Thereby, they gradually transfer from Design Thinking to software development when problem and solution space have been explored sufficiently.
- With the toolbox model, Design Thinking is introduced as a bundle of creative methods, which can be used to solve certain design problems at all stages of the software engineering process.

While these approaches have been identified through post-Design-Thinking-workshop interviews with people that work in the IT industry, a comprehensive evaluation of the success of these approaches or barriers and success factors for their application within the development processes of IT companies is yet to be done. Taking a look at actual industry projects, we found that Design Thinking is already being used for smaller software projects. IDEO, for example, developed IPhone applications targeted to young children and their families. The first project (IDEO 2009a) resulted in two games for young children that are intuitive and easy to play. In a later project (IDEO 2009b), IDEO created a series of Sesame Street -themed applications. The series includes a utility app for parents, which sends children to bed and entertains them, as well as several creative and educational apps for children. These examples show that Design Thinking can be used to successfully develop software products. However, creating software with the help of Design Thinking is not limited to design agencies like IDEO. Big software vendors like SAP or Microsoft are making efforts to integrate Design Thinking into their companies. During this year, SAP started creating so-called "app houses" around the world, where teams of "developer designers" experience a startup like process, resembling the integrated project model approach, to create small innovative apps for SAP (Savvas 2012). Creating subdivisions for innovation and research outside the actual company is a common approach and results in so-called innovation and research labs (Chartron et al. 2011). While this enables the labs to make use of processes and work flows that would not be used inside the main company, it also hinders the adoption of innovative processes in the rest of the company (Lindberg et al. 2011). Moreover, Brenner et al. found that this approach allows companies to be flexible and react quickly to changing demands but also results in an "IT of two speeds" (Brenner et al. 2012), separating the slow development in the main company and the fast and innovative development in the company's innovation divisions.

This separation also reflects the two degrees of complexity inherent in today's IT companies. Small-scale applications and research projects do not have to comply with all processes and restrictions that are present in large-scale enterprise software development. The development of large and complex enterprise software systems such as Enterprise Resource Planning (ERP), or Customer Relationship Management (CRM) systems is subject to numerous standards, legal constraints, and non-functional requirements, such as accessibility, availability and security (ISO/IEC 2005). To ensure compliance with these standards, software vendors create in-house quality assurance processes that every software product has to follow. Furthermore, larger software systems are developed in teams of hundreds or thousands of developers split into several smaller teams, which means that the chosen development processes must be scaled to accommodate this situation (cp. (Ambler 2009)). Thus the development of large enterprise applications poses unique problems that are not addressed by current efforts to integrate Design Thinking into software development, whereby Design Thinking is mainly implemented by smaller size projects and teams.

The reason for this could be that the large number of additional requirements and process restrictions, due to the complexity of the software, the standards, and the team sizes restrict the creativity and the freedom of thinking, which is necessary in Design Thinking. Another aspect might be that Design Thinking by default allows wild ideas that can be far away from what can be implemented in the company. Both are important aspects to consider when trying to integrate Design Thinking into enterprise software development processes. The existence of constructs such as innovation labs and app houses implies that innovation is desired by the industry but a process framework that integrates innovation methods like Design Thinking into large organizations is still to be found. SAP tried to bring Design Thinking into all their divisions by adopting an approach that resembles the split project model identified by Lindberg et al. (2012). As early as 2005, the company established a so-called Design Services Team (DST) whose purpose is "to improve the design of SAP software solutions as well as provide the organization with the means to scale up its adoption of Design Thinking" (Holloway 2009). Hildenbrand and Meyer took a different approach in a single project case study with SAP (Hildenbrand and Meyer 2012). They have successfully implemented a first prototypical process that follows the unified project model and integrates Design Thinking and Scrum into one process. They started their project with a Design Thinking phase that was followed by a development phase with Scrum and interleaved Design Thinking activities as needed. Hildenbrand and Meyer found that agile or lean development and Design Thinking have enough in common to allow a mixed mindset in their team and a smooth transition between the two. They identified user story maps as an especially helpful tool to bridge the gap between insights and ideas and actual product requirements during implementation. The case study provides a glimpse of what Design Thinking activities can achieve in software development. The proposed process, however, still needs to be further refined and tested with more project teams and different size projects. Furthermore, the process does not consider the additional requirements and restrictions that apply to enterprise software, an aspect that we believe to be critical for the success of Design Thinking in enterprise software development.

3 Research Agenda

As shown in the previous section, large software companies are making efforts to integrate Design Thinking into their development processes, but to a large degree the integration is taking place in smaller scale projects in innovation labs. With our research, we aim to identify currently existing approaches and methods for the integration of Design Thinking into software engineering processes within real companies and qualitatively assess their strengths and weaknesses. Based on these findings, we aim to create a prototypical development process framework that integrates Design Thinking activities into large-scale software development processes, such as scaled Scrum or Lean Development, while still respecting legal and organizational restrictions. This chapter provides an overview of our research strategies to achieve this goal.

3.1 Expert Interviews and Surveys

In order to identify currently existing Design Thinking approaches, we want to conduct expert interviews in software companies that are using or have used Design Thinking. Our target group for these interviews is software developers as well as employees from software vendors. They have tried, or were compelled to integrate Design Thinking into their development processes, having taken courses in Design Thinking or having been involved in external Design Thinking projects (e.g. as corporate liaisons in d.school or ME310 projects). Our goal is to analyze the impact of Design Thinking on development processes in those companies and its impact on personal work styles of the involved employees. Furthermore, we want to identify obstacles and opportunities for integrating Design Thinking into software companies. To this end, the following question will be investigated:

- Which companies or departments continued to use Design Thinking inside the company after initial test projects had ended? Which ones discontinued its adoption? What are the reasons for continuation or cancellation, respectively?
- If Design Thinking usage was continued, how was it applied?
    - Is it integrated into the development process or a complementary add-on, e.g. separate Design Thinking teams that consulted development teams?
    - Is it only implemented for certain departments and teams or applied on a broader scale?
    - Is it used voluntarily by only a few employees or is its use by a variety of employees actively encouraged (or enforced)?
    - Is it used as a set of creative tools (brainstorming, prototyping) or as a complete process, as, for example, described in (Plattner et al. 2009)?
- What are success factors and obstacles for the integration of Design Thinking in software companies?
    - How to achieve diversity in teams when working mainly with IT-specialists?
    - How to handle time and budget when adding Design Thinking activities to software engineering practices?
    - Are employees willing to accept the new methodologies or do they reject Design Thinking? How might we overcome the reasons for rejection?
- What methods and tools are applied to integrate Design Thinking into software development activities? Which might be desirable to support its adoption?
- Is Design Thinking helpful in all phases of software development processes or only during requirements engineering?

This is only an initial set of questions that will most likely be extended during the actual interviews. After the expert interviews, a survey targeting employees of software companies (e.g. software developers, product owners, requirement engineers, etc.) will be developed to quantitatively verify the findings and identify further possibilities for the integration of Design Thinking and software engineering processes.

3.2 Overview of Design Thinking Methods

Design Thinking offers a huge number of creative tools and methods that can help at different stages during the process. The following list provides a small excerpt of some of these methods and their usage during different stages of the Design Thinking process:

- Understand: 360 degree research, stakeholder map, sketches, low-fi prototypes (e.g. cardboard + paper)
- Observe: expert interviews, extreme users, surveys, walk in someone else's shoes, critical function prototype, critical experience prototype
- Synthesize: point of view statement, persona, scenario, process map, Venn diagram, two-by-two axis map
- Ideate: different brainstorming setups with the complete team or sub team (e.g. quiet the first 10 min then all together), brainstorming rules
- Prototype: lo-fi prototypes, hi-fi prototypes, dark horse prototypes, role plays, system prototype
- Test: interviews, extreme users, questionnaires, experiments

Currently we know of no existing overview of these methods or an evaluation of their applicability to software development. Therefore we aim at collecting methods, tools and frameworks used during Design Thinking activities in the before-mentioned expert interviews and surveys. We then want to create a comprehensive overview of the methods, their benefits and usage and map them to Design Thinking or development stages. This collection can then be used to create a "Design Thinking Toolbox" that helps teams to find the right method for their current need.

3.3 *Design Thinking and Scrum*

Similar to Hildenbrand and Meyer (Hildenbrand and Meyer 2012), we believe that Design Thinking and Agile Development can complement each other. Together with the Institute of Information Management at the University of St.Gallen,[6] we are currently working on a prototypical process that combines Design Thinking and Scrum. We follow the general idea of using Design Thinking to find, define, or refine requirements for the software in an initial project phase. In this phase, we aim to make Design Thinking activities projectable by planning them in form of sprints, providing a definition of done and estimating the effort of the planned tasks beforehand. The following development phase will follow a Scrum-like process with interwoven Design Thinking activities, such as user testing and storytelling, to keep the team and the development user centered. As this is only a first prototype, we will test and refine it to develop a process framework, which allows companies to fit/scale the process to their specific needs. The process will also serve as a basis to evaluate different team sizes and setups during software development with Design Thinking. Possible project setups include the approaches proposed by Lindberg et al. (2012):

[6] IWI HSG – http://www.iwi.unisg.ch/

- The project will be started with a specialized Design Thinking team during the initial phase and will later serve as the Product owner during development.
- The project will be started with a specialized Design Thinking team that will be gradually changed into the development team by dropping Design Thinking experts and integrating developers as needed.
- A team of developers trained in Design Thinking will work on the project the entire time.

In addition to these setups we will investigate the effects of guidance through Design Thinking coaches or a "Design Thinking master".

3.4 Integration of Non-functional Requirements and Quality Assurance

In order to integrate quality-assuring processes into the process model, we took a look at standard non-functional requirements for enterprise software and company processes that ensure the requirements are met. We found that these processes do not constantly check whether the requirements are fulfilled. Instead they require checks only at specific points during development. In most cases if these checks fail, for example if the performance is bad or code standards are not met, it will be a huge effort to go back and try to fix them. We think that constant feedback as to whether requirements are met or not would be of great help, because it would enable the development teams to react as soon as a problem emerges. We believe this is possible, because large parts of the non-functional requirements, such as certain usability, accessibility or performance requirements can be ensured via tool support. Those tools can then run at all times during the project, providing the feedback constantly. In our research we will investigate what tools can help to ensure performance, software quality, and other non-functional requirements and how they can be integrated into our process framework.

Naturally not all requirements can be checked with the help of software tools. Requirements concerning the documentation, the actual development process, the functional correctness or operations and support, as well as the actual usability of the developed software still need to be ensured by other means. Nevertheless these aspects will also be considered for our process framework and suitable methods to ensure the compliance with these requirements will be investigated.

With the help of the four described elements we aim to provide a comprehensive process framework that enables enterprise software vendors to integrate Design Thinking into their software development processes and provides the development teams with tools and methods to successfully use Design Thinking and make sure non-functional requirements are met. To evaluate our process framework as well as suitable teaching methods we intend to create a Design Thinking software development course, which is described in the following section.

4 An Educational Testbed for Integration Approaches

Based on the initial insights gathered through our interviews and analysis of related work, we aim at creating a prototypical project course that helps to assess different approaches for the integration of Design Thinking and software development processes. To minimize the impact of project failure within those experiments, they will be placed in an educational setting, first, before being repeated in industry settings.

Previous case studies have shown that teaching Design Thinking methodologies to software engineering teams poses various problems. Domschke et al. (2009) conclude that the common background and shared knowledge about methods and tools, as commonly found in homogenous teams, hinder the acceptance of new methodologies taught in Design Thinking education. They further observed that software engineers tend to refuse usage of methods and tools that do not yield technological outcomes but focus on intangible findings such as user needs. We want to overcome this gap by creating a project-based software engineering course based on the ME310 teaching approach. We henceforth name this course CS310.

4.1 Course Setup

Stanford University's Mechanical Engineering 310 (ME310) course has a long-standing track record as a case for research studies concerned with factors that influence the performance of design teams (Carleton and Leifer 2009). The course projects are collaborations between Stanford, international partner universities, and corporate partners. The interdisciplinary student teams tackle product innovation challenges that were proposed by the partner companies. Each project aims at delivering a mature and functional prototype to the corporate partners. In general, the projects are split into three phases (cp. Fig. 1). During the fall quarter, the teams are supposed to "make it up", that means exploring the problem space by performing extensive benchmarking, need finding, user observations, interviews, and creating initial prototypes. The second phase is about creating ideas for potential solutions and testing them in a series of prototypes with increasing fidelity. Finally, the spring quarter is dedicated to implementing the final prototype and only imposing minor adoptions to the design.

As the course is based in Stanford's mechanical engineering department, the main focus of most projects resides in the creation of physical prototypes. In recent years, however, an increasing number of projects have strongly focused on software solutions. As CS310 projects will solely focus on the development of software solutions, minor adoptions of the given ME310 course setup are necessary.

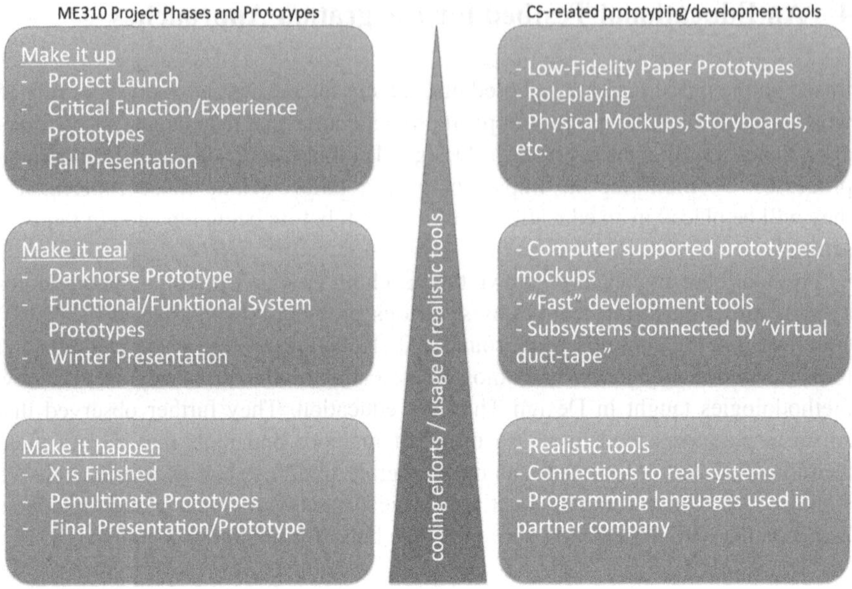

Fig. 1 Phases of the project course and their relation to employed prototyping tools and methods

4.2 *Adaptions of the ME310 Approach*

Conceptually, the three described phases will be kept within CS310 as well. Prototyping assignments, however, need to be adopted – especially in the second and third phase. Figure 1 outlines potential prototyping tools for the different phases. During the initial phases, ME310 and CS310 teams will most likely not differ vastly in their choice of tools. Fast and rapid prototyping with low-fidelity tools (paper, cardboard, Lego, role playing, etc.) is the primary working mode during this phase. Critical functions, however, potentially require some software development effort to realistically prove that certain functionality could be implemented given the prescribed technologies or systems.

Beginning with the second phase, the differences between projects of either course are increasing. Instead of creating functional system prototypes out of physical components, which at this point could still be connected by plain duct-tape, an analogous working mode needs to be found for software teams. Potential approaches could be the connection of computer supported mockups with briefly prototyped backend functionality or the manual connection of two separate software prototypes by transferring data manually between the systems instead of, for example, already programming the data transfer.

In the final phase, prototype building in CS310 becomes a pure software development effort. Using standard agile process models, such as Scrum, the students can structure their development efforts and plan ahead to deliver the required prototypes

"X is finished", "Penultimate", and, in the end, the final prototype. During this phase, it is crucial that realistic tools are being used to lower the transfer effort back to the corporate sponsors. If, for example, the teams would create their prototype using programming languages and frameworks that are not also used by their partner company, the entire prototype would have to be re-implemented after the project. If the teams are not familiar with the required tools or languages, additional training needs to be provided. Following the rationale for ME310's paper robot prototypes, an equivalent exercise could be used midway through the project to get the teams accustomed to the tools before the final phase begins.

The usage of realistic tools has another benefit: it provides a showcase for the corporate sponsor and outlines how they could apply the ME310/CS310 approach within their internal projects as well. This aspect of realism can be further fostered by applying some of the requirements presented in the related work section and lets the teams best implement them into the prototypes or at least provide concrete concepts for required integration efforts.

4.3 Future Directions

The first iteration of this course is performed using the integrated project approach, as it fits naturally into the project setup described above. An initial project was started in October 2012 and features collaboration between SAP, HPI, and the University of St. Gallen. This initial test project aims at evaluating both the overall teaching approach and initial approaches for the integration of software development processes and Design Thinking, as presented in the previous section. In future iterations of the course, we want to integrate a bigger number of teams to be able to compare different integration approaches with each other and/or try similar approaches in different project settings.

5 Evaluation

The ever increasing adoption of agile software development practices (Ambler 2007) is not solely driven by rigorous quantitative proof for their positive effects on independent variables that measure the success of a project. Instead, most startups employ the suitable methods rather naturally in an effort to avoid huge documentation overhead and to focus on rapidly creating usable increments with maximum customer benefit. Large companies initially test the approaches in smaller test projects and if these projects yield convincing results, a widespread adoption within the entire organization follows consequently. The same basically holds true for Design Thinking principles. Without having rigorous proof for the effectiveness of each method, companies employ them because they created satisfying results in previous projects and "simply feel right".

A primary goal of the HPI-Stanford Design Thinking Research Program, however, is to scientifically examine and measure the effects of Design Thinking principles on independent measures of success. Hence, if we are to conceive new process models that combine software development activities with Design Thinking, such proof is mandatory. This section outlines our intended evaluation process on two levels.

Firstly, we show how we use our analyzeD platform (Kowark and Plattner 2012) to collect and analyze artifacts of virtual team collaboration that are indicative of the adherence-to or negligence-of the principles we outlined, previously. Secondly, we briefly outline further data collection and analysis methods, such as questionnaires and interviews that are supposed to capture and quantify the effects of collaboration activities that are not observable within the digital collaboration tools. Following an action research approach (Baskerville 1999), we will collect all insights gained from each evaluation source and use them to conceive adoptions of our proposed process model and its methods. These adoptions will be tested again using the previously described methods, thus, creating the research cycle shown in Fig. 2.

5.1 Virtual Collaboration Analysis

In previous project years, we created a platform that has aimed to be a simple and readily accessible way of collecting and analyzing information about artifacts of virtual team collaboration, such as wiki pages, emails, bug tracking items, or source code revisions (Kowark et al. 2011a; Uflacker et al. 2010). These artifacts have proven to be effective in providing viable indicators for the adherence-to or negligence-of both Design Thinking and software engineering principles (Kowark and Plattner 2012; Uflacker 2010). In the following we outline how we plan to apply this tool and its recently added features within this research project.

5.1.1 AnalyzeD Usage Workflow

The workflow of virtual collaboration analysis with analyzeD is comprised of three basic steps:

1. Collection of data from the employed groupware systems,
2. Aggregation of data into network structures with optional anonymization,
3. Data analysis through graphical tools or by performing queries on the graph.

The first two steps can be performed with minimal manual intervention. Automated sensor clients collect usage information from the different groupware tools employed by the teams and transfers it to the analyzeD application. The data returned by the sensor clients is aggregated into so-called Team Collaboration

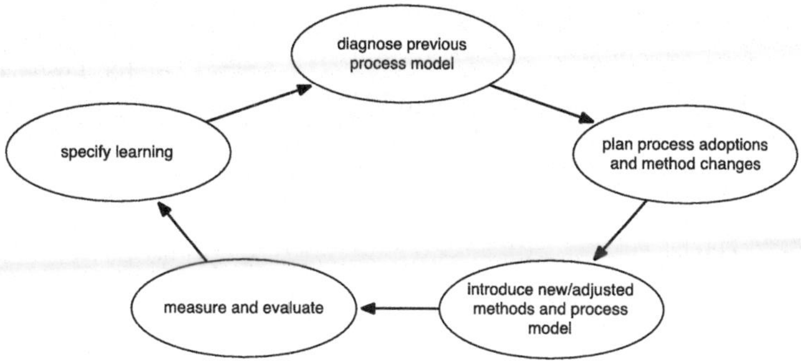

Fig. 2 Research Cycle used to continuously evaluate and improve the process model (Adopted from Baskerville 1999)

Networks (TCN) (Uflacker and Zeier 2010). These networks are built upon ontologies that represent concepts of different collaboration tools, such as wikis, email lists, or version control systems. All nodes can be linked to each other through relations and, thus, provide a comprehensive overview of observable virtual collaboration activities across tool boundaries.

Once the networks have been created, they can be analyzed in a multitude of ways. For example, standard social network analysis functionality incorporated into the underlying graph database[7] can be used to determine how closely people collaborate by counting the number of artifacts that they are jointly connected to. Through this and other analyses of the graphs, reoccurring sequences of collaboration behavior can be identified and their impact on process or product metrics, as captured along the collaboration activity, can be statistically evaluated.

In (Kowark et al. 2011a), we introduced a concept that allows generalizing such findings made within single projects and also detecting occurrences of similar behavior in other projects. In order to enable such cross-project analyses, we introduced the "Matcher" component to the general platform architecture (see Fig. 3). This component is able to take characteristics of the respective source and target networks into account and adopt the data queries accordingly. If, for example, a network pattern was initially observed in a project that uses Scrum as the development process, role names might differ from projects that use other process models. The Matcher component detects such differences and either uses previously defined equivalence relations or asks the users to provide new ones. Additionally, cardinalities and time constraints are translated to account for differences in project size and duration.

With the help of this mechanism, it is possible to transfer knowledge about potentially beneficial or detrimental virtual collaboration signatures between projects. Each user of the system can graphically model those signatures, categorize them to simplify later usage, and publish them in a central repository that is

[7] AllegroGraph – http://franz.com/agraph/allegrograph/

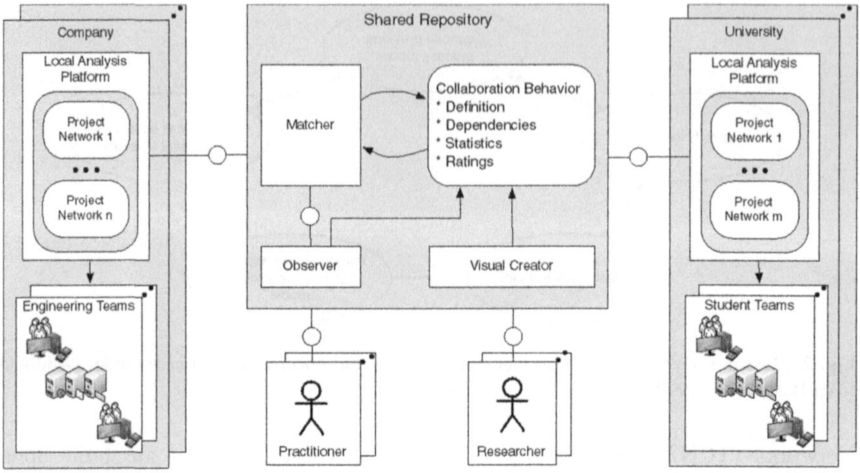

Fig. 3 AnalyzeD architecture overview showing the local analysis platforms that are connected with a central repository for virtual collaboration behavior

accessible by all platform users. Furthermore, it is possible to select patterns of interest and constantly monitor collaboration networks for newly detected occurrences. An example of this usage mode is presented in the following subsection. With regards to the analysis of virtual collaboration within our ME310-like computer science project course, we will use this approach to model various assumptions about the student's collaboration behavior and determine whether they correlate with collected metrics, such as development velocity of the teams (Schwaber and Beedle 2001), independently collected ratings of their work, or self-reported team satisfaction. Initial questions that could be answered by our evaluation approach are:

- When did teams start to write source code and move away from physical prototypes? Does an earlier or later start have an impact on the project outcome?
- Which prototype source code was transferred into the final prototype and which and how much code was discarded during the process?
- How many people worked on the same code? Does the number of programmers affect the performance of the final outcome?

5.1.2 Example Application

The following section provides a practical example for the possibilities of virtual collaboration analysis. It is based on a software engineering course at HPI (Kowark et al. 2011b). In this course, 6–8 student teams jointly develop an enterprise system. They are provided with an extensive groupware infrastructure and Scrum is prescribed as their development process. Being an agile software development process,

Scrum, however, does not focus on tools but on individuals and interactions. Hence, while we provide them with the groupware tools, the teams are supposed to find the most suitable solutions for their work instead of dogmatically adhering to existing systems.

During post-hoc analysis of the first iterations of the course, we made the following observations: when teams started to focus on physical Scrum boards for ticket maintenance, the digital versions of the tickets were increasingly updated in bulk by designated members of the development teams. This permits maintaining an overview of the current progress directly within the team office. At the same time it makes it possible to share the progress with other teams and the teaching staff through the groupware system with minimal overhead. In order to detect this working behavior in subsequent projects, we used the previously described modeling approach to model two contrary graph patterns.

Figure 4 depicts a ticket being changed by its owner. The owner here is not the person that created the ticket – which would often be the Product Owner – but the person who was assigned to implement the described feature or perform the desired task. We assumed that such behavior would mostly be exploited at the beginning of the project and decrease over time when small changes are carried out exclusively on the physical Scrum board.

The second behavior – updates by other team members that do not own the ticket – is depicted in Fig. 5. It is assumed to increase during the project as more changes are performed on the physical Scrum board and designated team members start to update the digital tickets every couple of days for their teammates. Figure 6 shows the detected number of occurrences during the project for an entire development group. It is clear that the assumptions about the occurrence rate of the behaviors generally held true. After many developers initially edited their own tickets on a regular basis, turnaround was reached midway through the project, where changes by non-owners became more frequent.

Despite their seemingly strong indication towards a real change in the working behavior of the teams, virtual collaboration analysis can only provide circumstantial evidence and should be accompanied by intensive manual observations. Firstly, digital collaboration constitutes only a fraction of overall collaboration activity. Secondly, if the subjects under investigation know that their virtual collaboration behavior is being analyzed, observer effects might come into play, with people (un) consciously trying to generate "favorable" digital signatures. Hence, such analyses should be used mainly as a project management aid that helps to detect potentially detrimental or beneficial behavior and serve as a starting point for further, more targeted manual investigations. We will use analyzeD in such a manner and with the help of the "Visual Creator" for graph sequences, we are enabled to frequently model and adapt assumptions about the students' working behavior and test their influences on collected process and product metrics.

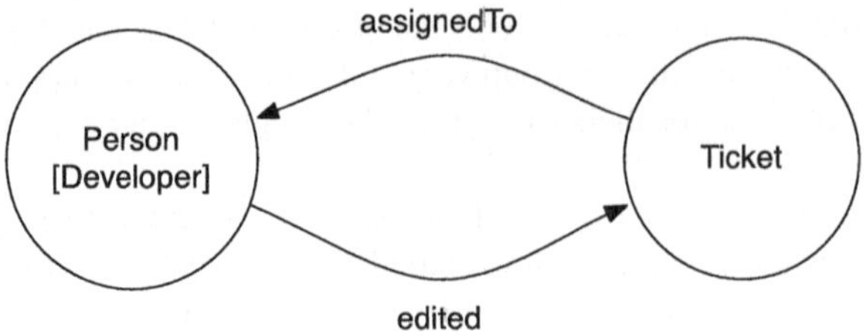

Fig. 4 Behavior 1. A ticket is changed by the developer that owns it

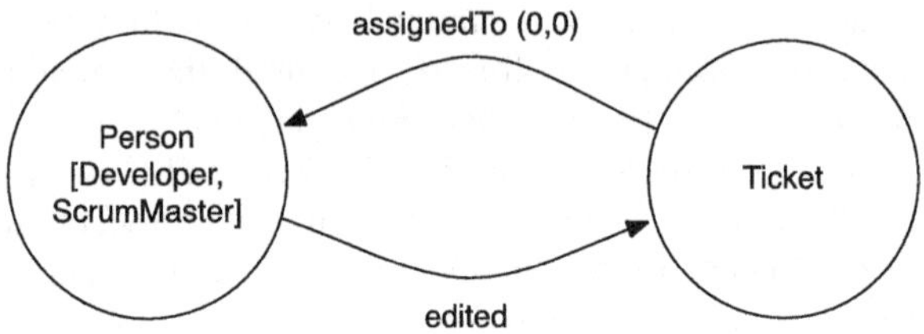

Fig. 5 Behavior 2. A ticket is changed by a team member that is not the owner of the ticket

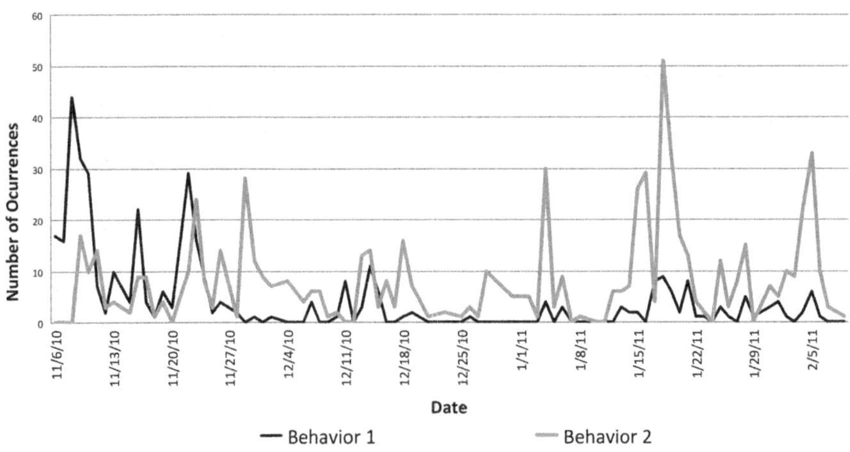

Fig. 6 Development of occurrences of the behaviors under investigation over the course of the project

5.1.3 Content Analysis

The previous examples were solely focused on the structural analysis of team collaboration networks. In a recent master's thesis, we also investigated possible applications for an analysis of node contents. Specifically, the thesis used latent Dirichlet Allocation (LDA) (Blei et al. 2003), a statistical model, to automatically

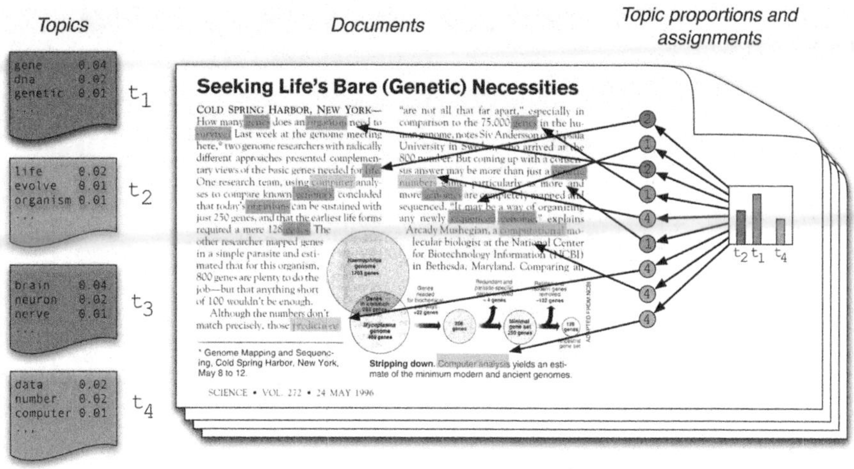

Fig. 7 The generative process of LDA (Adapted from Blei et al. 2003)

determine which topics are addressed by the different collaboration artifacts. The algorithm takes a given input document (e.g., email bodies, source code revision commit messages, wiki page contents) and detects potentially contained topics through an analysis of word frequency (see Fig. 7). For a complete description of the algorithm, the interested reader is referred to the original thesis (Gehrer 2012).

Building on the possibilities that topic detection in collaboration artefacts offers, we created a prototypical topic explorer component for analyzeD (see Fig. 8 for screenshot). This component allows filtering the plethora of available collaboration artifacts to only those that exhibit a chosen topic. Furthermore, the implementation is able to provide a heatmap for the chosen topics. This heatmap, which is shown on the top left corner of the screenshot, allows the possibility of identifying during which phases of the project certain topics were "hot" and when their importance decreases. This could equally help to determine which ideas within a project were carried out from initial prototyping phases to the final implementation and which were discarded along the way of the process.

The histogram in the top right corner displays the distribution of a topic amongst the different collaboration channels. In the example shown, it is possible to see that implementation efforts (marked yellow) only started after initial documentation and planning efforts (emails, tickets, wiki pages – red, green, and blue). This functionality further aids the structural analysis described in the previous section, as we are not only able to see when the project members started implementations in general but can break this down to a per-topic-level.

Finally, as the TCN link artifacts to their creators or editors, the additional knowledge about exhibited topics makes it possible to determine potential experts for certain topics. The "Experts" area in the lower left corner prototypically shows

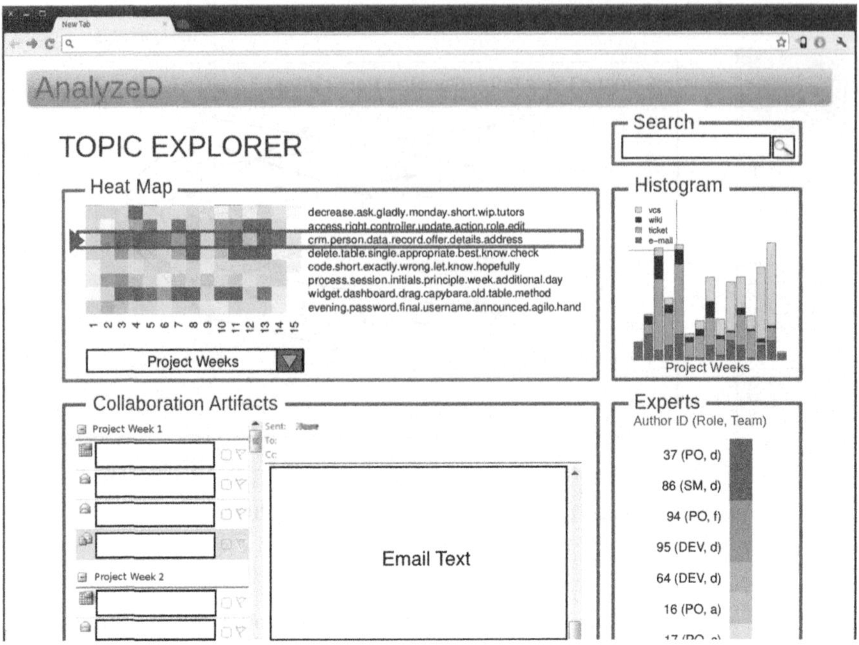

Fig. 8 Screenshot of the topic explorer prototype

how it is easily possible to determine that Person 37 and 86 potentially are most knowledgeable about the given topic or at least might be good initial contact persons.

5.2 *Manual Data Collection*

In addition to analyzing virtual collaboration activities, a variety of manual data collection efforts will take place throughout the projects. Firstly, we will ask the project members to constantly evaluate their own work. Questionnaires will record metrics, such as team satisfaction or subjective perception of the team's progress, but also provide feedback about the used tools and methods. Secondly, external experts will further rate the prototypes and the final outcome of the projects in terms of innovation, desirability for the customer, and difficulty of the initial problem. In addition to the external evaluation, we are also interested in the teams self assessment as well as the assessment of teaching assistants and tutors to investigate not only the project outcomes but also assess which parts of the development process and the integrated Design Thinking activities were considered useful and which did not aid the project. Based on these ratings and the analysis of virtual collaboration activities, we will constantly refine our proposed process model and aim at providing a set of indicators that allow monitoring whether the teams comply with it or if intervention might become necessary.

6 Conclusion

We believe that an integration of Design Thinking into agile software development processes can benefit the development of large-scale enterprise software. Therefore we want to create a process framework that integrates tools and methods to comply with legal restrictions and non-functional requirements that are common in enterprise software development together with Design Thinking and agile methodologies. To achieve this goal, we first aim to identify success factors and obstacles for the adoption of Design Thinking in IT companies by interviewing company employees that received d.school training, as well as employees that were involved in previous and ongoing attempts to integrate Design Thinking into software development processes. Based on their feedback, we aim to conceive a prototypical software development process that integrates Design Thinking practices with agile software engineering activities. We also take a look at existing quality assurance processes, tools, standards and legal requirements for enterprise software development to integrate them into our own process model.

The process will be tested in a software engineering course that is based on ME310 but adopted to the specifics of software engineering. By integrating the aforementioned interview partners as liaisons and coaches we want to further open up a direct feedback channel to industry that helps them to overcome the previously identified obstacles with the help of means that were evaluated within the course. Furthermore, we use our previously created software platform analyzeD to evaluate our process and the teams in our course. The feedback received from the industry, our teams and our evaluations will be used to constantly refine and improve the course and our process model.

References

Ambler SW (2007) Survey Says. Agile has crossed the chasm. Dr Dobbs J 32(8):59–61

Ambler SW (2009) The agile scaling model (ASM): adapting agile methods for complex environments. Retrieved from ftp://ftp.software.ibm.com/common/ssi/sa/wh/n/raw14204usen/RAW14204USEN.PDF

Anderson DJ (2010) Kanban: successful evolutionary change for your technology business. Blue Hole Press, Sequim

Baskerville RL (1999) Investigating information systems with action research. Commun AIS 2 (3es):4

Beckman SL, Barry M (2007) Innovation as a learning process: embedding design thinking. Calif Manage Rev 50(1):25–56

Blei DM, Ng AY, Jordan MI (2003) Latent dirichlet allocation. J Mach Learn Res 3:993–1022. Retrieved from http://dl.acm.org/citation.cfm?id=944937

Brenner W, Uebernickel F, Wulf J, Zelt S, Győry A, Heym M, Warnke A (2012) Strategies for application management services. Retrieved from http://www.navisco.com/download/casestudies/navisco_strategies_ams.pdf

BusinessWeek (2004) The Power of Design, {BusinessWeek} Cover Story. BusinessWeek

Carleton T, Leifer L (2009) Stanford's {ME310} course as an evolution of engineering design. In: Proceedings of the 19th CIRP design conference–competitive design. Cranfield, United Kingdom

Chartron J, Steinhoff DF, Paulssen PDM (2011) User Driven Innovation@Telekom Laboratories – Das Innovationsforum. Ideenmanagement 1-2011, 14–17

Domschke M, Bog A, Zeier A (2009) Teaching design thinking to software engineers: two future-oriented curriculum case studies. In: 26th ICSID World Design Congress. Singapore

Gehrer R (2012, July). Extraction of emerging topics from digital collaboration artifacts of software engineering teams. Hasso Plattner Institute for IT Systems Engineering

Hildenbrand T, Meyer J (2012) Intertwining lean and design thinking: software product development from empathy to shipment. In: Maedche A, Botzenhardt A, Neer L (eds) Software for people. Springer Berlin Heidelberg, pp 217–237

Holloway M (2009) How tangible is your strategy? How design thinking can turn your strategy into reality. J Bus Strat 30(2/3):50–56

IDEO (2009a) iPhone Applications textbar IDEO. Retrieved from http://www.ideo.com/work/iphone-applications

IDEO (2009b) Sesame Street iPhone Apps IDEO. Retrieved from http://www.ideo.com/work/sesame-street-iphone-apps/

ISO/IEC (2005) {ISO/IEC} 25000:2005 – Software engineering – Software product quality requirements and evaluation (SQuaRE) – Guide to SQuaRE

Kowark T, Dobrigkeit P, Zeier A (2011a) Towards a shared repository for patterns in virtual team collaboration. In: 5th international conference on new trends in information science and service science. Macao, China

Kowark T, Müller J, Müller S, Zeier A (2011b) An educational testbed for the computational analysis of collaboration in early stages of software development processes. In: Proceedings of the 44th Hawaii international conference on system sciences (HICSS). Koloa, Kauai, HI, USA

Kowark T, Plattner H (2012) AnalyzeD: a shared tool for analyzing virtual team collaboration in classroom software engineering projects. In: The 2012 international conference on frontiers in education: computer science and computer engineering. Las Vegas, NV, USA

Lindberg T, Meinel C, Wagner R (2011) Design thinking: a fruitful concept for IT development? In: Meinel C, Leifer L, Plattner H (eds) Design thinking research: studying Co-creation in practice. Springer Berlin Heidelberg, pp 3–18

Lindberg T, Köppen E, Rauth I, Meinel C (2012) On the perception, adoption and implementation of design thinking in the IT industry. Des Think Res 229–240

Lund A (2011) March 11: a case study of applying design thinking at Microsoft Corporation Human Centered Design & Engineering. Retrieved from http://www.hcde.washington.edu/nav-courses/521/win11/mar11

Plattner H, Meinel C, Weinberg U (2009) Design thinking. Mi-Verlag, Munich

Savvas A (2012, April) SAP focuses on product "desirability" through design to lure customers – ComputerworldUK.com. Retrieved from http://www.computerworlduk.com/news/applications/3402011/sap-focuses-on-product-desirability-through-design-lure-customers/

Schwaber K, Beedle M (2001) Agile software development with Scrum. Prentice Hall PTR, Upper Saddle River

Thomke S, Feinberg B (2010) Design thinking and innovation at Apple. Harv Bus Sch 9-609-066

Uflacker M (2010) Monitoring virtual team collaboration: methods, applications, and experiences in engineering design. Hasso Plattner Institute for IT Systems Engineering, Potsdam

Uflacker M, Zeier A (2010) A semantic network approach to analyzing virtual team interactions in the early stages of conceptual design. Future Gener Comput Sys 27(1):88–99

Uflacker M, Kowark T, Zeier A (2010) An instrument for real-time design interaction capture and analysis. In: Meinel C, Leifer L (eds) Design thinking: understand – improve – apply. Springer (in print)

Sharing Knowledge Through Tangible Models: Designing Kickoff Workshops for Agile Software Development Projects

Markus Guentert, Alexander Luebbe, and Mathias Weske

Abstract In software engineering, the programmer depends on precise descriptions of the system to be built. To get these descriptions, analysts condense the knowledge about the domain from observations and discussions with the users, the people that will eventually work with the software. The users have to communicate their knowledge about the domain and express their needs. With TBPM we have shown that it is possible for end users to express themselves by means of process models. We now transfer these findings to other fields in software engineering. We investigated in the discipline of requirements engineering, especially in the context of agile software development approaches. From practitioners we learned that during the first iterations, code tends to be thrown away completely since the initial requirements gathering phase is intentionally kept lean. We therefore introduced the concept of *need-finding iterations* and tackle this problem in our research. We develop a holistic workshop methodology to kick off agile software development projects in which a shared understanding among stakeholders is to be fostered. Discussions that would arise after a *software prototype* has been implemented are encouraged to be conducted at an earlier stage by making use of an adequate modeling solution. We propose *story prototypes* which essentially enrich user stories with control flow information and thereby are enhanced to show the big picture rather than just individual aspects of the system to be built. In such a kickoff workshop we encourage a detailed need-finding together with the customer by means of shared model building.

M. Guentert (✉) • A. Luebbe • M. Weske
Hasso Plattner Institute at the University of Potsdam, Potsdam, Germany
e-mail: markus.guentert@hpi.uni-potsdam.de

H. Plattner et al. (eds.), *Design Thinking Research*, Understanding Innovation,
DOI 10.1007/978-3-319-01303-9_13,

1 Introduction

There is a saying that to really build a great product, you have to build something that you love to use yourself (Fried and Hansson 2010). This approach brought about great innovations in consumer electronics and also software systems. Products such as Google Maps, Twitter, Facebook or Delicious were created by programmers to satisfy their personal needs. In these cases, the programmer has the knowledge about what is desirable and what is technically feasible. Unfortunately, for the vast majority of software programs this setting is not given. Business software, written to support the operations of a company, is by definition no consumer product. Thus, the programmer will never be the user of software that does risk management, supply-chain management or accounting. The programmer has the knowledge about technical feasibility, but the desirable functionality originates from another person, the user of the system.

To enable software engineers to understand the requirements of the user, a plethora of approaches was created in the last decades in software engineering. As one example, the research area of requirements engineering (e.g. Alexander and Stevens 2002; Davis et al. 2006) investigates how information is derived from the user, documented and maintained towards the implementation of a software program.

In the Design Thinking Research Program previous projects have investigated paths for better requirements engineering. For example, the project by Prof. Giese (PI) on scenario-based prototyping (Gabrysiak et al. 2009; Gabrysiak et al. 2011), enabled the end user to simulate daily scenarios, in a role playing game to collect information about the workflows and associated data. This information was then used to enrich and validate requirements. The results are data models and behavioral models that are checked for plausibility before they go to the further software engineering process (Fig. 1).

In a previous project, we investigated a new approach to joint model building together with the end users. In the case of Tangible Business Process Modeling (TBPM) (Edelman et al. 2009; Luebbe and Weske 2011), business processes are the activities performed in organizations. They are represented in graphical models to enable better communication about them. They are key artifacts for the software engineers as they also describe how people and IT systems interlink. Typically, the end users communicate their knowledge about the process verbally. An analyst afterwards represents this knowledge as a business process model. In the project on TBPM the process model is co-created by the end users and the analyst acts as a facilitator. He explains the concepts behind process modeling and the semantics of the artifacts on the table. The resulting models are then used in the further software engineering process. In his dissertation, Alexander Luebbe (Luebbe 2011) created a methodology to involve domain experts more strongly in business process modeling and has shown that it leads to more engaged end users and better validated models.

Fig. 1 People creating a business process model together using TBPM, a set of tangible shapes and whiteboard markers. A modeling expert facilitates the session

2 Transferring the Findings of TBPM

TBPM enables process analysts to co-create modeling artifacts together with end users who have no modeling experience upfront. This collective modeling facilitates and sharpens communication between the process participant and the process analyst. As of now we have applied our methodology to the field of process elicitation. It is our intention to also transfer the findings of the TBPM project to other fields in software engineering, especially those which heavily rely on modeling artifacts.

In verbal communication, information is likely to get lost or misinterpreted (Lin et al. 2005). To ease this problem, software engineers have created modeling languages, such as BPMN (OMG 2011) for business process modeling. It defines a set of concepts, their relations and how they are visually represented. For example, BPMN defines the concepts of activities that are in an ordered relationship to each other. An activity is defined as a chunk of work that takes time to be performed and is assigned to one responsible role. A role is another concept that summarizes a set of abilities and, e.g. a clerk.

Typically such information is gathered by a specially trained analyst. The end users, experts in the domain, can provide this information about their work environment. The model is then created afterwards by the analyst. The TBPM project has shown that it is possible to enable end users to directly create the process models that are required for the software engineering project. Therefore, a tangible incarnation of BPMN (OMG 2011), the process notation, was created and evaluated. In a workshop, the analyst teaches the core concepts of process modeling, the semantics of the shapes and facilitates the creation of the process models. In that case, he can integrate the views of the people at the table into one overall agreed model. The research on TBPM suggests (Luebbe 2011) that the people learn more about the process when they model it together and they also iterate the process model more often. Given these positive effects, one might ask: Does this idea also work for other software engineering models?

On the most general level, the models that describe software systems can be divided into behavioral models, data models and architectural models. Behavioral models describe the dynamic in the system of which the software is part of. For example, a business process is a behavioral model. It may contain human steps as well as steps taken inside the software system. A data model describes how information is structured. For example, an Entity-Relationship Diagrams (Chen 1976) describes the information attributes and their relationship. More concrete, an address is an information attribute of a person. An address consists of street name, postcode and country. Finally, architectural models describe which logical components exist within a software system and how they interrelate.

Despite the three fundamental types of models in software engineering, there exist a countless number of modeling languages developed in the history of computer science to describe aspects of a software system. Each of these modeling languages has been developed for a special purpose. Some modeling notations have been developed by software companies as a unique selling point (e.g. Keller et al. 1992). Others were developed by researchers to fill a particular need (e.g. van der Aalst and ter Hofstede 2005). And again other modeling notations have been developed by standardization committees (e.g. OMG 2011). These languages are made for different audiences and show a different level of technical detail. Furthermore, they embody a different set of modeling concepts even if their core nature is of the same type. For example, the aforementioned process modeling notation BPMN (OMG 2011) is a behavioral model, that also has the notion of data-based decisions and system communication through systems collaborating using message exchange. Thus, they have basic concepts from data modeling and architectural modeling but at their core they focus on behavioral models. Activity diagrams (Dumas and ter Hofstede 2001), also a type of behavioral models, are designed to describe internal system behavior and have no idea of interacting systems. Thus, even within one fundamental modeling type there is a huge variety of modeling languages. Barely any language is purely of only one type. But even the same concepts are depicted differently, in different notations. To unify the plethora of modeling notations, the unified modeling language (Fowler and Scott 2000) was introduced in the 1990s. It defines 13+ modeling notations for the purpose of software engineering. Despite its adoption in teaching (Zeichick 2004), UML has not unified the scattered modeling world in the software engineering industry. Moreover, there are also alternative sets of modeling languages that claim to cover all necessary aspects of software engineering (e.g. Gausemeier et al. 2009) (Fig. 2).

Picking a modeling language or even a set of modeling languages is very difficult because they are hard to compare. For business process models, a special type of behavioral model, this was done based on their capabilities to express frequently used modeling situations (Wohed et al. 2006; Russell et al. 2006). But one could also argue that popularity should be the key to choosing the language. Another option was offered by Moody (Moody 2009), who developed a metric to measure the aesthetics of modeling languages based on the physical attributes of the notation. It allows identifying more or less suitable languages based on objective measures that he validated through experiments.

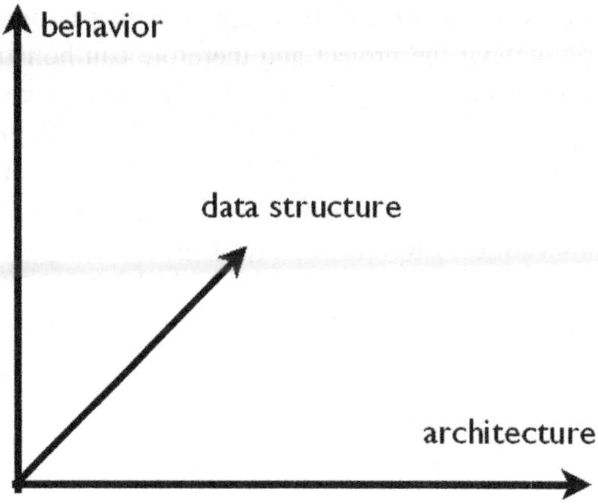

Fig. 2 Fundamental dimensions of modeling languages. Each modeling language focuses on one of these dimensions but typically is a mix of many concepts describing behavior, data structure and architecture

We believe that potentially every one of the mentioned modeling languages could be incarnated as a tangible toolset comparable with TBPM. Simply creating such toolsets, however, is not our research interest. We instead look into other scenarios which could make use of the methodology behind TBPM, which essentially is to involve end users in the creation of such models.

3 The Changing Role of Models in Requirements Engineering Over Time

The field of software engineering has established a systematic approach to gather requirements from the customer. Such a *requirements engineering* phase is traditionally incarnated at the beginning of the project, i.e., before the implementation starts. In traditional requirements engineering – as especially reflected in the Waterfall model (Royce 1970) – dedicated requirements engineers elicit the desired functionality of the software to be implemented together with the customer. This results in a heavyweight process which involves different stakeholders over a long period of time. As an outcome of this process, requirements engineers create a specification document which consolidates the knowledge accumulated from the customer. Such a specification document most often contains multiple 100 pages and is regarded as a static artifact in the further progress of the process. The specification document, including a variety of software engineering models, serves as primary input source for the implementation. Those who implement the desired software are usually not the same people as the requirements engineers. Therefore, the developers may hardly ever be in direct contact with the customer and need to rely on the information provided in the specification.

As the field developed, major drawbacks have been identified in the strictly linear sequence of requirements engineering and the implementation phase. At first, the customer of the software to be implemented is unlikely to be aware of all the requirements at the beginning, i.e., before a concrete prototype of the software exists. A common misconception is that requirements are considered to be stable throughout the project and therefore can holistically be elicited in an early stage. Requirements may already be outdated once the implementation phase starts. Furthermore the customer may not be able to precisely formulate the requirements and to state them in an unambiguous way. A verbal discussion about software requirements is of an abstract character. Since the specification document and, i.e., the formal models, are created without the presence of the customer, the information exchanged reflects an individual interpretation by the requirements engineers. This is a common cause for misunderstandings. In addition, the validation of requirements is non-trivial since the resulting models and their formalism may not be understood by the customer.

In order to overcome these drawbacks, *agile* software development approaches have entered the stage. Rather than following a strictly linear approach between requirements gathering and implementation, these two phases are alternating in fixed length time units. Whereas misunderstandings in requirements communication, and the accompanying wrong software outputs that occur in the traditional approach, are revealed at a very late stage, i.e., after implementation, agile approaches intend to clarify such misunderstandings early on. When later misunderstandings are revealed and more code has been produced, the more expensive it will become to introduce necessary changes to the system (Pohl 2010). (Boehm and Basili 2005) states that the effort of fixing a conceptual error in a software product after delivery is up to 100 times higher than during the requirements engineering phase.

Agile software development approaches focus on continuously producing outputs, i.e., *software prototypes*, which foster feedback from the customer. Thereby requirements can be validated or neglected on a much more concrete basis. By demoing a software prototype to the customer, he is better able to comprehend the requirements in context and their implications rather than in theoretical discussions. Each feedback cycle is used to revise the requirements step-by-step rather than defining them all-at-once. An intense requirements engineering upfront the implementation is therefore replaced by the iterative character of implementing a set of requirements one at a time. Furthermore, detailed specification documents hereby become obsolete.

User stories are commonly used to communicate requirements with customers. Originally evolved from the Extreme Programming approach (Steinberg and Palmer 2003), the concept of user stories has been adapted by a significant amount of agile practitioners nowadays. A user story is ideally one sentence of natural language with a fixed structure which describes a desired software requirement intended for a specific user role. The structure follows the form:

As a <*user role*> I want to <*goal*> so that <*benefit*>.

The key point is that by means of user stories, requirements are formulated from the perspective of the user rather than from the developer. Both this fact and the short wording enable the customer to directly participate in the creation and discussion of software requirements. User stories are a means to conserve the discussion about a requirement, but never to holistically specify it since this would be contrary to the concept of short wording. Therefore in order to understand a user story, a stakeholder should ideally have participated in the discussion around it.

User stories serve as primary *models* in the development process. As with TBPM, they can directly be communicated with the customer and the customer is able to express himself by way of these models. User stories can be *aggregated* to a common subject, i.e., to a superordinate functionality, or *decomposed* if one is found to be too general and that there are multiple stories behind it. User stories, however, only reflect the conceptual view from the perspective of the respective user role.

4 Do User Stories Tell the Whole Story?

In semi-structured interviews with agile practitioners (N = 5, among them two product owners, two Scrum Masters and one lecturer in the field of Requirements Engineering), we have learned that besides user stories other types of software engineering models, which serve as communication artifacts, are created additionally. Especially activity diagrams (OMG 2012), which essentially are process models, are being created among developers to map the system processes of the software to be implemented. The customer is not involved in the creation of them since these more technical models are believed to not be understood by the customer. The customer is therefore not able to provide feedback as to whether the content reflected in such diagrams has been understood correctly.

User stories function by expressing a particular aspect of the system. Once a few user stories have been discussed and interpreted, different aspects of the software to be implemented are reflected. What is lacking, however, is the overall context, i.e., the interrelation between user stories. The collection of gathered user stories can initially be regarded as a simple list with a flat hierarchy. In order to overcome this flat hierarchy, user stories are aggregated and decomposed as described above. Both the aggregation and decomposition, however, solely group user stories for the purpose of categorization. The context of one user story is hereby indicated only by other related user stories. Additionally a group name is provided to enrich the context information.

What remains unclear, however, is the sequence between user stories, i.e., the order in which they may be executed. Imagine the two following user stories had been elicited: “As a logistics clerk I want to check the order status so that I can follow up whether the customer paid for the goods.” (A) and “As a logistics clerk I want to initiate the shipping process so that the customer will receive his goods.” (B). It may be of significant difference for the resulting software whether user story A needs to be completed in the system process before user story B can be started. It is also imaginable the other way around that user story B is always executed before user story A. As a third alternative user stories A and B could be treated independently and their order relationship may be of no importance.

Another important concept for the context of user stories is decisions. Consider the following user stories: "As a logistics clerk I want to print a shipping label so that the goods can be handed over to the post office." (C) and "As a logistics clerk I want to cancel the order if the goods are not available." (D). It is valuable information whether C and D are both executed or if just one is performed. The relationship can be more complex also, e.g., first C is executed and then – depending on the result of C – D may possibly be executed, but D is never executed before C. This type of information is not captured in user stories.

5 User Stories from a Process Perspective: Story Prototypes

The elaborated concepts bring in the *process* perspective. We see a great potential to enrich a collection of user stories with this additional information. This results in a sharpening of the communication between developers and customers. We propose depicting system processes by means of user stories and their order relationship, furthermore considering the concepts of decisions and independence. Requirements can thereby be specified more precisely without introducing a significant overhead of other model types.

Since user stories are formulated in natural language, their underlying concept and mechanism can be communicated with the customer in a rather straightforward and effortless fashion. No technical pre-knowledge is necessary to understand the formalism and after the customer has had the chance to become familiar with the mode of thought, he will soon gain confidence to directly participate in their creation. The experience of agile practitioners confirms this notion.

With TBPM, referring to both the toolset and methodology, we have shown, on the other hand, that it is also possible for end users, i.e. domain experts, to express their knowledge by means of process models (Luebbe and Weske 2011). If the concepts of process modeling are introduced iteratively, that is step-by-step rather than all at once, end users with no former experience about the "process perspective" are able to follow the discussion and even create their own process models.

Seeing the potential that both model types are potentially being understood by the customer, we believe that it is reasonable to combine user stories with concepts from process modeling. We thereby create additional meaning and context information to software requirements but are still able to communicate them between developers and the customer. Baring in mind that agile software development approaches continuously produce software prototypes as means for discussion, we call our resulting models *story prototypes*. We intend to prepone the discussion that arises by demoing a software prototype to co-created story prototypes in a stage prior to implementation.

6 Minimizing Need-Finding Iterations

Essential to the background of agile software development approaches is the concept of early feedback. In order to avoid misunderstandings and thereby the resulting wrong development tracks, the customer is exposed to concrete software prototypes continuously from the beginning. In this way, requirements can be validated and rethought since they appear in a concrete context, i.e., a running software demo. Agile practitioners therefore commonly believe that it is inappropriate to keep discussions about software requirements on a theoretical level, as has been practiced in traditional requirements engineering for a long time. Feedback on concrete software prototypes replaces theoretical discussion. If a software prototype turns out to not meet the intended requirements, it is still considered to be of value since this fact itself has been revealed.. By means of iteration, the prototypes can be refined continuously.

The implementation of software prototypes, however, is costly and time-intensive. A common duration for a development cycle, e.g. in Scrum (Schwaber 2004) projects, is 2 weeks. During this time the development resources are bound and dedicated to it. Feedback on the result is available only after these 2 weeks. Especially in the first iterations of a project when requirements are still vague and imprecise, it is imaginable that the implemented software prototypes do not meet the customer's expectations.

In interviews with agile practitioners (see above) we learned that the first iterations of an agile software development project are primarily used for a detailed need-finding. The software prototypes produced support this process. Since requirements are likely to be revised noticeably in this stage, the code that has been produced for the software prototypes tends to be thrown out afterwards.

We intend to tackle this factor and reduce the amount of such *need-finding iterations* in which code is being produced but thrown away afterwards. We believe that parts of the discussion which arises with the demo of a software prototype can be conducted in an earlier stage in the project, i.e., before any implementation is done at all. We believe that such discussions can also be facilitated by an adequate utilization of models. Thus misunderstandings shall be clarified even earlier on and a shared understanding fostered among the different stakeholders.

7 A New Scenario: Kickoff Workshops

How are IT projects actually started nowadays? In many cases a *kickoff workshop* is organized where the key stakeholders meet (Smith and Reinertsen 1998). Also the literature on agile software development approaches proposes holding an initial requirements gathering workshop (Cohn 2004). But what exactly should happen during such a workshop? The pertinent literature mentions few techniques and provides little methodology.

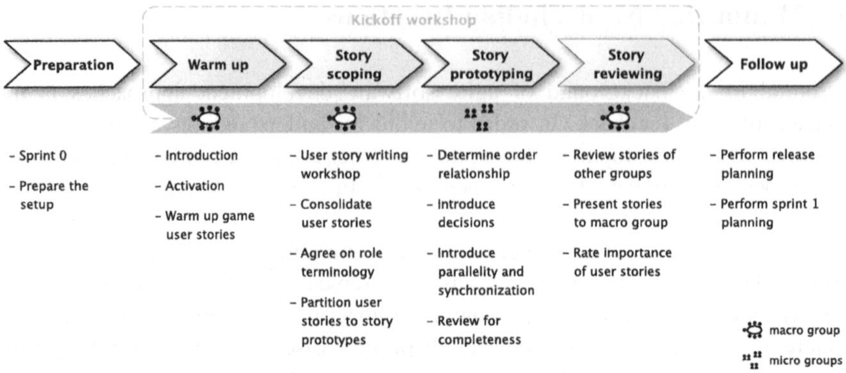

Fig. 3 Kickoff workshop procedure. *Macro group* refers to activities carried out by all workshop participants together and *micro groups* to activities distributed to smaller teams according to a vertical decomposition of the system to be built

We see an important potential to use the kickoff workshop to sharpen the communication especially between the customer and the developers in the project. The kickoff workshop determines the tenor of the project and is a great opportunity for an IT service provider to leave a competent impression and to motivate the customer for the project. Furthermore, the more the different stakeholders' opinions are integrated, the more satisfied they might approach the further progress of the project. In our research, we focus on designing a holistic workshop methodology to kick off agile software development projects. We consider the preparation as well as the facilitation of the workshop. The kickoff workshop procedure is illustrated in Fig. 3 and elaborated hereafter.

In advance of the kickoff workshop it is reasonable to consider a so-called "Sprint Zero" (Soni 2011), which is held without the customer being present. It may be used for team-building activities among the developers, to set up the IT infrastructure, the workspace and to gather initial thoughts on the domain in preparation for the project. Optionally, a small demo project may be developed with the intended technology to be used throughout the project in order to become familiar with the development environment. Such a Sprint 0 should not use more than 1 week of time.

An integral part of the preparation of the workshop itself is to prepare the setup and to set up the right atmosphere. From design thinking research we learn that room and time are both crucial factors which could compromise effectiveness and productivity (Broß et al. 2011). The room should ideally provide enough space for the participants to sit, stand and walk around to allow different perspectives on both the room and the subject. A flexible room layout is therefore conducive to a good result. Time can be used most efficiently if the time unit constraints for each task are communicated. It is good practice to communicate the entire schedule at the beginning and to clarify whether participants need to be absent at some point of time. We consider a workshop duration of 1 or 2 days as suitable. Compacting the workshop into entire working days is recommended, otherwise the overhead of reactivating the group to explain concepts and mechanisms again increases.

As soon as the participants of the workshop arrive, a traditional round of introduction is made. Every participant should also express his expectations of the project and the workshop. In total this should not exceed 3 min per participant. This conservative approach is meant to address the attendees' anticipation of the workshop. After the introduction, the moderator communicates the timetable and provides instructions for the "activation phase". The activation phase can be shaped in a variety of forms but should ideally include a short physical activity. This should get the participants to stand up and explore the setting. In this way, they get familiar with the room and come closer to other participants but also prepare for the active attitude they will need to work. Furthermore, they lose the fear of contact and get accustomed to speaking and contributing to the group. This can be regarded as an initial form of team-building.

After the activation phase, initial thoughts are exchanged on the software to be implemented. The idea is to conduct a user story writing workshop as proposed in (Cohn 2004). As a preparation, the moderator explains the concept of user stories and presents examples to the group. In order to familiarize the group with this thinking mode, a short warm up game is arranged in which the participants actively contribute to the creation of user stories. Concepts can be discussed during this activity, without putting actual content at stake. It furthermore increases the confidence of the participants in expressing themselves. The scenario of the warm up game should be largely unrelated to the actual project and characterize a well-known setting familiar to everybody in the group. The positive value of warm up games has been discovered in the TBPM context. Scenarios proposed for TBPM were withdrawing money from an ATM or home-ordering pizza from a pizza service.

The story writing workshop should at first have a different character of brainstorming. Every participant is therefore instructed to silently collect the requirements that come to mind in the form of user stories. This collection can be mainly unclassified and is intended to share ideas from different perspectives. Individual silent brainstorming avoids the confrontation of different personalities, e.g., a dominant versus a reserved, personality, and therefore ensures that every participant has contributed his ideas to the group (Diehl and Stroebe 1987). Ideally each user story is written on a separate Post-It note. After participants are done with this task, each participant briefly presents his user stories to the others. If another participant wrote down the same or a very similar story, she can directly add her Post-It note to the other one. After this first consolidation phase, key user roles can be identified from the existing user stories. It is important to invest time in a discussion involving the wording of user roles since they may later be used to represent personas.

As a next step, user stories are divided into story prototypes which reflect independent components of the system. Therefore, dependent user stories are aggregated into groups. The concept of dependence should be oriented toward the process perspective, i.e., user stories should be aggregated which occur in the same process context. The different system processes are thereby revealed and the corresponding story prototypes are scoped. Furthermore, the user roles involved in each system process become visible. The results are identified story prototypes with an initial collection of associated requirements, together with role information.

Until now all activities have been carried out by the entire group of participants together. We refer to this group as the *macro group*. Now the macro group is divided into smaller teams – so called *micro groups* – to examine the identified system processes in detail, whereby every team works out one particular story prototype and presents it to the macro group afterwards. This approach follows the concept of vertical decomposition. For each story prototype, if necessary, overlapping user stories are grouped and consolidated. The resulting stories are then transcribed to the tangible toolset (each story on one tangible shape) and laid out on a table, like in the TBPM methodology. First their order relationship is roughly determined. If two user stories are to be executed one after the other, they are placed next to each other. This results in a linear sketch of the story prototype.

After a rough process layout has been prototyped, the concept of decisions is introduced. Thereby information about exclusiveness is enriched to the linear stream of requirements so far. OR-Gateways as found in the tangible toolset are hence incorporated into the process and the paths of user stories rearranged accordingly. As a next step the concept of parallel execution and synchronization is explained. If in one and the same story prototype, user stories may be executed in parallel, i.e., their execution order is independent, this shall be reflected by AND-gateways of the tangible toolset. While reviewing the system process it may be revealed that additional user stories, which have so far not been considered, need to be reflected in the story prototype. By making the context transparent,, this enriches the model in terms of completeness. The results of this phase are detailed story prototypes complemented by control flow concepts.

The worked out story prototypes can now optionally be reviewed by other teams and corrected if necessary. Each team should briefly present its results to the macro group. As a final step for the workshop the identified requirements are prioritized along the created models. High priority features become candidates for the first implementation iteration to be prototyped and demoed afterwards. As a follow up to the kickoff workshop an initial release planning is performed. Furthermore, the first implementation sprint needs to be scoped. All relevant information should be present as a result of the workshop. The actual planning, however, might in practice be undertaken without the presence of the customer.

8 Benefits of a Kickoff Workshop

One could argue that creating models as a preparation for implementation brings back the thought-pattern established in traditional requirements engineering, whereby requirements were meant to be fixed in the beginning. However, considering potential need-finding iterations in agile development projects, it seems reasonable to devote time to thinking about how things are approached if an adequate and effective tool for the discussion is available.

The key benefits of our model framework and workshop technique is that all participants are directly involved in the creation of results and thereby contribute to the shaping and scoping of the project. The models are co-created and thereby help the stakeholders to identify with the results. The interactive character of the workshop fosters an active attitude to work and overcomes hierarchies. By applying the model framework, user stories are enriched by context information which strengthens the understanding of the whole picture rather than just starting with a collection of different system aspects. The time needed to gather this big picture view of the system to be built is small compared to an entire requirements engineering phase. The tradeoff between added value and time to introduce, or apply the additional concepts respectively, is favorable for our approach. We make use of a light-weight addition which delivers direct benefits in a shared understanding among the stakeholders. The kickoff workshop furthermore represents a good opportunity for the stakeholders to get to know each other in the proposed group work assignments. It can additionally be used to break in the rituals that are to be incorporated in the project, i.e., different methods proposed by agile software development approaches.

Such a kickoff strategy may be applied to projects in which a new software system is to be implemented for one customer or potentially many customers. It is important to have at least one customer being integrated to the project so that the development team is provided with real-life instead of just theoretical requirements. If the software to be implemented, on the other hand, is not essentially new but rather a custom development, it is a good idea to consider one or more reference systems. These can be used to identify what parts can be taken over, but also how the desired system differs. The duration of a kickoff workshop is certainly context-specific and depends on the number of stakeholders, as well as on the complexity of the project. We recommend a kickoff workshop of no more than 2 days of duration because participants might otherwise get lost in details and complexity. This would not only contrast to the agile way of thinking, but also resemble the drawbacks that have surfaced in traditional requirements engineering.

9 Conclusion

The TBPM methodology allows us to facilitate a co-creation of process models with novice users, i.e., domain experts. It is thereby possible to sharpen otherwise informal communication to a very precise formalism. The setting encourages group discussions and is perceived to be more fun and engaging than traditional interview methods. The primary objective of our research is to transfer the findings from TBPM to other fields in software engineering as well. After an initially systematic approach in which we concentrated on other types of commonly used models, we realized that potentially every language may be incarnated in a "tangible" fashion.

This, however, did not reveal much opportunity for research. We have instead investigated other scenarios which might profit from the methodology behind TBPM.

Agile software development approaches are gaining more and more meaning nowadays. This led us to look at the role that models play in this context. Requirements are primarily communicated by means of user stories and are intended to be implemented in software prototypes that can be discussed with the customer. An initial capturing of requirements is intentionally kept lean. After interviewing agile practitioners, we identified the concept of need-finding iterations, that is, initial implementation cycles which are solely carried out to foster a better understanding of the problem, with the actual code thrown away afterwards.

We tackle this problem by proposing a kickoff workshop in which the discussion that arises with the demoing of a software prototype is conducted prior to co-creation of models. We therefore introduce story prototypes which essentially are user stories enriched with context information that reflect the process perspective behind it. A simple collection of user stories only expresses certain aspects of the system to be built, but not their inter-relation. Rather than creating additional models without the presence of the customer, we make use of a tangible toolset, lay out the user stories on a table and enrich them with control flow concepts like decisions and parallelism. Our contribution therefore is two-sided: On the one hand, we introduce a kickoff workshop methodology which compiles different best practices of group work techniques and the co-creation of models. The output of the workshop, on the other hand, results in a new type of model which we call story prototype.

In the next steps of our research we intend to develop project characteristics from which we derive a set of metrics that allows us to quantify need-finding iterations. We can then relate the desired productivity increase that arises by application of our kickoff workshop technique to reference projects not making use of the technique. We will collaborate with an industry partner currently following a suitable agile development approach and gather data from "regular" projects. Then we incarnate kickoff workshops in upcoming projects and evaluate the results against the data from these reference projects.

References

Alexander IF, Stevens R (2002) Writing better requirements. Addison-Wesley, London

Boehm B, Basili V (2005) Software defect reduction top 10 list. In: Foundations of empirical software engineering: the legacy of Victor R. Basili, p 426

Broß J, Noweski C, Meinel C (2011) Reviving the innovative process of design thinking. In: ICIW 2011, The 6th international conference on internet and web applications and services, pp 142–149

Chen PP (1976) The entity-relationship model – toward a unified view of data. ACM Trans Database Syst 1(1):9–36

Cohn M (2004) User stories applied: for agile software development. Addison-Wesley, Boston
Davis A et al (2006) Effectiveness of requirements elicitation techniques: empirical results derived from a systematic review. In: 14th IEEE international conference requirements engineering, pp 179–188
Diehl M, Stroebe W (1987) Productivity loss in brainstorming groups: toward the solution of a riddle. J Pers Soc Psychol 53(3):497
Dumas M, ter Hofstede A (2001) UML activity diagrams as a workflow specification language. In: The unified modeling language. Modeling languages, concepts, and tools, S.76–90
Edelman J, Grosskopf A, Weske M (2009) Tangible business process modeling: a new approach. In: Proceedings of the 17th international conference on engineering design, ICED'09
Fowler M, Scott K (2000) UML distilled: a brief guide to the standard object modeling language. Addison-Wesley, Reading
Fried J, Hansson DH (2010) ReWork: change the way you work forever, Ebury
Gabrysiak G, Giese H, Seibel A (2009) Interactive visualization for elicitation and validationn of requirements with scenario-based prototyping. In: Requirements engineering visualization (REV), 2009 4th international workshop on, pp 41–45
Gabrysiak G, Giese H, Seibel A (2011) Towards next generation design thinking: scenario-based prototyping for designing complex software systems with multiple users. Des Think 219–236
Gausemeier J, Plass C, Wenzelmann C (2009) Zukunftsorientierte Unternehmensgestaltung—Strategien, Geschäftsprozesse und IT-Systeme für die Produktion von morgen, Hanser Fachbuch
Grosskopf A, Weske M (2010) On business process model reviews. In: Workshop proceedings of ER-POIS: empirical research on process oriented information systems affiliated to CAiSE10, pp 31–42
Keller G, Nüttgens M, Scheer A-W (1992) Semantische Prozessmodellierung auf der Grundlage "Ereignisgesteuerter Prozessketten (EPK)". Institut für Wirtschaftsinformatik, Saarbrücken
Laue R, Gadatsch A (2010) Measuring the understandability of business process models – are we asking the right questions? In: Proceedings of the 6th international workshop on business process design (BPD 2010)
Lin L, Geng X, Whinston AB (2005) A sender-receiver framework for knowledge transfer. MIS Quart 29:197–219
Luebbe A (2011) Tangible business process modeling – design and evaluation of a process model elicitation technique. Ph.D. dissertation, Hasso Plattner institute for IT systems engineering, University of Potsdam
Luebbe A, Weske M (2011) Determining the effect of tangible business process modeling. In: Plattner H, Meinel C, Leifer LJ (eds) Design thinking – studying co-creation in practice. Springer, New York
Martin RC (2003) Agile software development: principles, patterns, and practices. Prentice Hall, Upper Saddle River
Moody DL (2009) The "physics" of notations: toward a scientific basis for constructing visual notations in software engineering. IEEE Trans Software Eng 35:756–779
Object Management Group (2011) Business process model and notation (BPMN) 2.0
Object Management Group (2012) Unified modeling language (UML) 2.5
Patig S (2008) A practical guide to testing the understandability of notations. In: Proceedings of the 5th on Asia-Pacific conference on conceptual modelling, vol 79, pp 49–58
Pohl K (2010) Requirements engineering: fundamentals, principles, and techniques. Springer, New York
Recker J, Dreiling A (2009) Does it really matter which process modeling grammar we use? An experimental study on understanding process models. Inform Software Tech
Royce W (1970) Managing the development of large software systems. In: Proceedings of IEEE WESCON, vol 26. Los Angeles

Russell N et al (2006) On the suitability of UML 2.0 activity diagrams for business process modelling. In: Proceedings of the 3rd Asia-Pacific conference on conceptual modelling, vol 53, pp 95–104

Schwaber K (2004) Agile project management with Scrum. Microsoft Press, Redmond

Smith P, Reinertsen D (1998) Developing products in half the time: new rules, new tools. Wiley, New York

Soni N, Soni A (2011) Agile release planning. Arete Solutions LLC

Steinberg DH, Palmer DW (2003) Extreme software engineering a hands-on approach. Prentice-Hall, Upper Saddle River

Van Der Aalst WMP, Ter Hofstede AHM (2005) YAWL: yet another workflow language. Inf Syst 30(4):245–275

Wohed P et al (2006) On the suitability of bpmn for business process modelling. Lect Notes Comput Sci 4102:161

Zeichick A (2004) UML adoption making strong progress. Software development times, 15 Aug 2004

How to Compare Performance in Program Design Activities: Towards an Empirical Evaluation of CoExist

Bastian Steinert and Robert Hirschfeld

Abstract We present the design of an empirical experiment to compare programmers' performance in program design tasks. The experiment is targeted to empirically examine the benefits of CoExist, a set of extensions to programming environments. CoExist supports programmers in dealing with unexpected and undesired consequences of making changes to their code base. Changing source code involves the risk of making errors. For example, a promising idea to simplify the code can suddenly turn out inappropriate, a situation that, if not prepared, requires programmers to manually withdraw recent changes. Traditionally, programmers have to strictly follow a structured and disciplined approach to reduce the costs of making errors. However, this traditional approach requires planning for upcoming but still uncertain changes in advance, which is time-consuming and also error prone. In addition, it requires significant effort to not forget the regular execution of the required activities, in particular in situations full of uncertainty. In contrast to this, CoExist offers dedicated tool support to recover fast and easily from undesired consequences. We believe that the presence of such tools encourages programmers to make source code changes at the moment they think of them, independent of whether or not the implications of such changes are already apparent. The presented experiment design to compare performance in program design tasks will help to examine this hypothesis.

1 Introduction

Programming involves more than continuously adding new lines of source code to a program. It requires reasoning about already written source code and making changes to it. Change is, for example, necessary to implement newly identified functionality or to fix erroneous behavior. Since programmers know that tomorrow

B. Steinert (✉) • R. Hirschfeld
Software Architecture Group, Hasso Plattner Institute, University of Potsdam, Potsdam, Germany
e-mail: bastian.steinert@hpi.uni-potsdam.de; robert.hirschfeld@hpi.uni-potsdam.de

H. Plattner et al. (eds.), *Design Thinking Research*, Understanding Innovation, DOI 10.1007/978-3-319-01303-9_14, they will have to make changes to the source code written today, they spend time on structuring it well and making it easy to understand, so that the modifications of tomorrow are easier to accomplish. In addition to the tasks of tomorrow, programmers also restructure their source code to ease the implementation of features of current interest (Beck and Andres 2004; Beck 1996; Fowler 1999).

Changing source code, however, involves risks because programmers can make errors. A promising idea to simplify the code can suddenly turn out inappropriate later in the process. Also, programmers can have flaws in their reasoning or can unintentionally ignore relevant aspects. Making an error represents a risk because it typically requires compensational activities that are time consuming and tedious to carry out.

The recommended way to deal with the risks of change is to anticipate that errors will be made and to continuously perform activities that keep compensation costs low. This includes checking for errors early and often as well as maintaining a safety net of stable development states one can fall back to (Beck and Andres 2004; Apache Software 2009). However, such precautionary activities can only reduce but not avoid the risk of tedious recovery work. Moreover, they can easily be ignored and distract from the actual task at hand.

We have developed an alternative way to deal with the risks involved in changing source code (Steinert et al. 2012). In contrast to asking for manual risk and cost reduction, we propose the provision of tool support such as CoExist that helps programmers deal with undesired consequences of their work. CoExist offers dedicated support to withdraw changes, to recover knowledge from previous development states, or to locate the cause of program failures even late in the process. In this way, CoExist avoids that making an error implies tedious recovery.

We believe that such a tool-based approach is preferable over a manual method based approach, which requires continuously following a set of practices. Research findings on design and cognition suggest that externalizing ideas supports the exploration of the problem and possible solutions. For example, creating prototypes help pursue a line of thought and discover unforeseen implications (Goldschmidt 1991; Lim et al. 2008). Other findings suggest that the creation of external representations inspires new associations (Kirsh 2010; Suwa and Tversky 2002). CoExist enables programmers to make changes as they think of them, because making errors does not imply additional costs.

We have designed an experiment to empirically examine our hypothesis that CoExist better supports programmers in program design activities, particularly in situations full of uncertainty. We have decided on a two by two mixed groups factorial design. We repeatedly measure subject performance in two different tasks, thereby having two groups of subjects, one using CoExist in addition to the regular tools for the second task. We measure performance by coding the changes and quantifying the effort for each category of change. In this report, we contribute the design and its justification for an experiment to empirically examine programmers' performance in program design activities.

Given a list of numbers: 3, 6, 1, 9, 10 ..., find all numbers smaller than 5.

```
givenNumbers:= "... a list of numbers ..."
result := OrderedCollection new.
1 to: givenNumbers size do: [:i | | eachNumber |
          eachNumber := givenNumbers at: i.
          eachNumber < 5
               ifTrue: [result add: eachNumber]]
```

```
givenNumbers := "... a list of numbers ..."
result := givenNumbers select: [:each | each < 5]
```

Fig. 1 Two different program snippets (in pseudo code) to fulfill the above stated need

2 Why Programming Involves Change or the Need for Well-Designed Programs

Programmers care about program qualities besides completeness and correctness. They care about a program's layout, its structure, or in which way a certain sub-goal is expressed. To illustrate this point, Fig. 1 shows two program snippets that lead to same observable effect (behavior) but are expressed in very different ways.

In the above example, as is often the case with design, there is hardly a right or wrong solution. Instead, the different solutions in the design space can be more or less appropriate for different purposes. While the first alternative includes more details about how the computation of the desired effect is accomplished, which can be important in certain domains, the second alternative focuses more on what should be achieved. It is also more concise. Programmers care about such design aspects because they know these will affect future programming activities. Appropriate source code design helps in two ways: it eases the implementation of new functionality and it supports maintenance and evolution.

2.1 How Source Code Design Affects Current Implementation Tasks

The amount of programming effort needed to implement a particular part of the problem depends on what has already been implemented. To illustrate this point, we build on the above shown program snippets (Fig. 1). The second solution can be so concise because some programmer has previously defined the program construct `select:`. And due to the availability of this construct, only little code is needed for programming needs similar to selecting a subset of numbers as shown in Fig. 2.

The decomposition of programs into modules helps to manage complexity (Blackwell 2002). It makes it possible to only be specific about details relevant and inherent to a particular concept, and to leave out the details of others. A proper

```
"Given a list of names, find all names matching 'jones'"
givenNames select: [:each | each matches: 'jones']
"Given a list of dates, find all dates before today"
givenDates select: [:each | each isBefore: Date today]
```

Fig. 2 Using `select:` to achieve similar needs in a concise manner

decomposition eases creating and comprehending the individual parts, and thus supports work on large complex systems (Parnas 1972).

However, how to decompose a program properly is not apparent from the beginning. The qualities of the current decomposition are revealed during the work on the program. The implementation of a particular aspect can be simple and straight- forward or rather complicated. The resulting source code can appear concise and right to the point or it can appear lengthy and disordered. Thus programmers often contemplate the source code and check whether it is easy enough to understand or still unnecessarily complex. They try to find elegant and simple solutions to express the concerns of the problem. Thereby, they can always introduce new modules (or programming constructs), such as selecting, or refining existing ones to better fit the current needs.

2.2 How Source Code Design Affects Program Maintenance and Evolution

The other reason why source code design matters is because programmers will have to (re-) understand formerly written source code and will have to adapt it. It is probably hard to find a piece of code that once written has never been changed afterwards. Programmers have to work on source code written by others as well as on source code they have written for some time. In both cases, they have to gain an in-depth understanding of the source code, either because they have never seen it before or because they cannot remember it in sufficient detail. In any case, the design of the source code can either facilitate or impede gaining a proper understanding of its meaning and effects.

The reason that programmers revisit previously written source code is because programming is a process of learning. It is not the same as assembling a car engine from a set of pre-defined parts, and it is also different from constructing a house according to a blueprint. Typically, every program is the first of its kind and the client does not have any type of a blueprint nor a meaningful description of the problem domain. Quite the contrary, programmers face a rather unstructured and little defined problem domain, which is not surprising considering that clients typically do not have the need to reason about their domain in a level of detail required for computer programs. In addition, the set of requirements often changes faster than the entire software solution can be accomplished.

To deal with these characteristics, software development often follows an incremental and iterative approach to eventually deal with all relevant aspects of the problem domain and to satisfy all needs. In an example development scenario, a client and the programmers sit together to identify the most important features for a small first version. The identified manageable set of features is realized without thinking much about further needs. Programmers may consult the client whenever questions arise during development. After releasing version 1, the client and the programmers talk about the desired functionality for version 2.

But for version 2, the client might want to extend the functionality implemented for version 1, which requires adapting and modifying the existing code to fulfill the extended requirements. Furthermore, the domain concepts implemented in successive program versions often depend on each other. It is likely that implementing functionality for a particular version builds on domain concepts that have already been implemented for former versions. But the existing code might only be partially sufficient and thus also requires adaptation or enhancement.

In addition to this macro level of software development, where clients and programmers collaborate to find out what needs to be built and in which order, programmers also have to deal with complexity and uncertainty on a micro level. Here, programmers are mainly concerned with how to built the defined features. Therefore, they follow a similar iterative and incremental approach. Programmers focus on a particular aspect of the problem, implement it, and thereafter consider another aspect, which possibly requires changing the code written for previous aspects.

On every level, the understanding of the problem domain co-evolves with the implementation (Dorst and Cross 2001). Seeing a first version of a solution improves the understanding of what the problem actually is and how it should be solved. However, as the problem understanding improves, so does the implementation. Thus, programmers will revisit previously written code and refine or adapt it to the improved understanding.

3 Why Changing Programs Involves Risks

While programmers regularly make changes to their programs, changing programs always involves the risk of creating *errors*. We distinguish *errors* from *mistakes*.

> ... a mistake is usually caused by poor judgment or a disregard of [known and understood] rules or principles, while an error implies an unintentional deviation from standards of accuracy or right conduct ... (Lindberg 2008)

Making an error refers to a situation where a programmers believes in the appropriateness of current and planned actions, and only later, after seeing the results, recognizes unexpected and undesired consequences. Making an error represents a risk because it often requires the programmer to accomplish some kind of tedious work to recover from it.

3.1 Risk Experienced, an Illustrative Story

To illustrate how changing programs involves risks, we would like to report the experience of a master's student. The student had been working on a visualization task using the Qt framework. At some point, he recognized that he had added several methods that all work on the same data. He decided to extract a class dedicated to this data structure and these methods. He created a new class called `SemanticLens` as a subclass of a Qt class `QEllipse`.

After moving all methods in this new class and adapting his code to make proper use of the new class, he contemplated his code and became skeptical about the decision to subclass the Qt class. He remembered that subclassing has the drawback of exposing the interface of the superclass to all clients who might make use of it, thereby creating a dependency that can become difficult during maintenance tasks. So, he decided to go for the delegation pattern instead. He was sure that delegation is the right way to go. So, the student changed the superclass of the `SemanticLense` class, added a field and accessor-methods to maintain a reference to a `QEllipse` object and also added initialization code. He changed the methods in `SemanticLense` class and made the required changes in the code using this class.

However, while looking at the result of making all these changes, the student realized that his assumption had been wrong. He could now see that subclassing is preferable over delegation in this situation because having access to the methods of the super-class is actually useful in his program. As a consequence of this insight, the student now faced the laborious task of manually withdrawing all the changes previously made to replace subclassing with delegation. He had to identify the relevant artifacts (files), and for each file to apply the undo command an undefined number of times until reaching the desired state. Such tasks are not only time-consuming but also tedious. Assuming that the changes made for the initial replacement had taken several minutes, this meant that manually withdrawing also took a few minutes. The required recovery work would have been even more laborious if the student had made further changes before recognizing the error. Such situations easily lead to irritation and are those programmers want to avoid.

One might argue that the student behaved incorrectly in this situation. He could have finished implementing the functionality first before considering refactoring the code. However, this could have led to more code needing adaptation later on. One could also argue that the student could have made a local commit (a checkpoint) before starting the refactoring, so that there would have been an easy way back. However, the code was in an immediate state and his work was not yet finished. He also has the very typical habit of thinking about commits only on completion of a task.

One could also argue, that he should have thought more carefully about his ideas before making any changes. Analyzing his idea in more depth might have been enough to raise doubts. However, it is unclear how much deliberation is necessary to avoid such errors. Moreover, too much thinking and questioning can easily lead to counterproductivity.

In contrast to the idea of careful upfront thinking, research findings on design and cognition suggest that externalizing ideas supports the exploration of the problem and possible solutions. For example, creating prototypes help pursue a line of thought and discover unforeseen implications (Goldschmidt 1991; Lim et al. 2008). Other findings suggest that the creation of external representations inspires new associations (Kirsh 2010; Suwa and Tversky 2002). While these results suggest that programmers should support their thinking by doing, this will inevitably imply the constitution of errors.

3.2 The Risks of Change and Methods to Reduce It

The issues in changing source code are broader and more general than illustrated by the story of the master's student balancing the pros and cons of delegation and subclassing. Writing and changing source code always involves the following risks:

- A promising idea unexpectedly turns out inappropriate and the programmer wants to continue exploring a previous idea, but many parts of the source code have already been modified.
- The improvement to one part of the program seems to affect the overall program behavior in unexpected ways, but the programmer has difficulties to find out which of the recent changes is causing the undesired behavior.
- The source code under current improvement turns out to be more complex than the programmer expected and it is unclear how the code was previously working.
- Recent changes turn out to represent multiple independent improvements that should be shared in separate increments.

The above listed issues are all well known. Literature describes them in detail, teachers tell students about them, and every programmer has experienced them (and still does) in some form or another. To reduce the risk of encountering such situations, literature recommends following a structured and disciplined approach and employing certain practices of work, which are, for example:

- To only work on one thing at a time, specifically a task or issue that you understand in detail. Therefore, to break larger task items down into smaller ones. This encourages staying on track and simplifies sharing your improvements with others.
- Make sure you understand the items you work on, if not, consider breaking down items into manageable parts, or consider consciously deciding on a phase of experimenting that should be preceded by committing (and/or branching).
- Write tests and run them often and regularly, at best, after every small change. This helps recognize bugs early in the process, and helps to pin down the cause of the problem to a few recent changes.
- Employ a distributed version control system such as *Git* or *Mercurial* and make frequent use of it by regularly committing small increments, which enables going back to a stable state easily.

The general pattern of these practices follows the contention that programmers should anticipate that they will make errors and thus should perform precautionary activities continuously and regularly in order to keep possible recovery costs low.

However, as the story about the student from above illustrates, the recommended way of relying on best practices does not seem sufficient in all circumstances. There are multiple reasons why this approach can fail in avoiding a need for tedious recovery activities:

- Wrong assessment. As was the case with the student in the story above, programmers can wrongly assess a programming situation. Applying best practices requires interpretation and subjective judgment. While running tests *often* or working on only *one thing* at a time are recommended, these are only guiding rules, and the programmer has to decide what *often* means and what the appropriate granularity of *things* is. Whenever a programmer feels confident about an idea and is unaware of any risks, the possibility remains that the assessment is wrong. In this case, the programmer will be unprepared for errors and will have to recover from them.
- Additional workload. Applying best practices is a continuous effort in addition to the work on the problem domain. As such, it is easy to forget and to ignore, in particular when one is caught up in creativity. Furthermore, mustering the needed discipline and remembering "not to forget" require significant mental effort (Allen 2001), even so practice helps reduce the required amount of attention.
- Upfront thinking. Following the recommended path requires programmers to structure the work ahead of them. This is a consequence of the need for interpretation and value judgment. For example, the practice to work on only one thing at a time requires regular reflection about current and upcoming work and assessing whether it should still be considered as "one thing" (a logical unit of work). Becoming aware of potential future risks requires thinking about the situation without working on it.

These limitations show that programmers will arrive, every now and then, in a situation where they have to face tedious work to recover from a previously made error, like the student in the story above.

4 CoExist: Tools to Encourage Change by Avoiding Risks

We have developed CoExist, an extension to the programming environment Squeak/Smalltalk, to preserve previous development states and to provide immediate access to relevant information (Steinert et al. 2012). CoExist is based on the key insight that the risks are caused by the loss of information during the process of change. With every change, we lose a previous version, unless it is saved explicitly. This version, however, can be of value in future development states, when, for example, an idea turns out inappropriate.

The basis of CoExist takes care of preserving potentially valuable information. It continuously performs commits in the background. Every change to the code base leads to a new version one can go back to. To make the user aware of this background

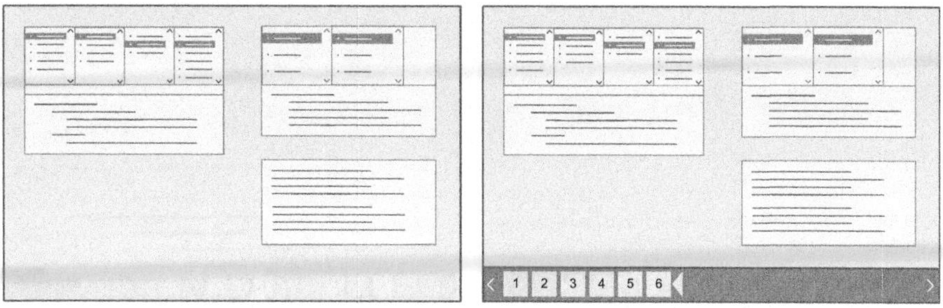

Fig. 3 The user interface of a regular Squeak/Smalltalk programming environment (on the *left*), and the version bar element added in CoExist (on the *right*)

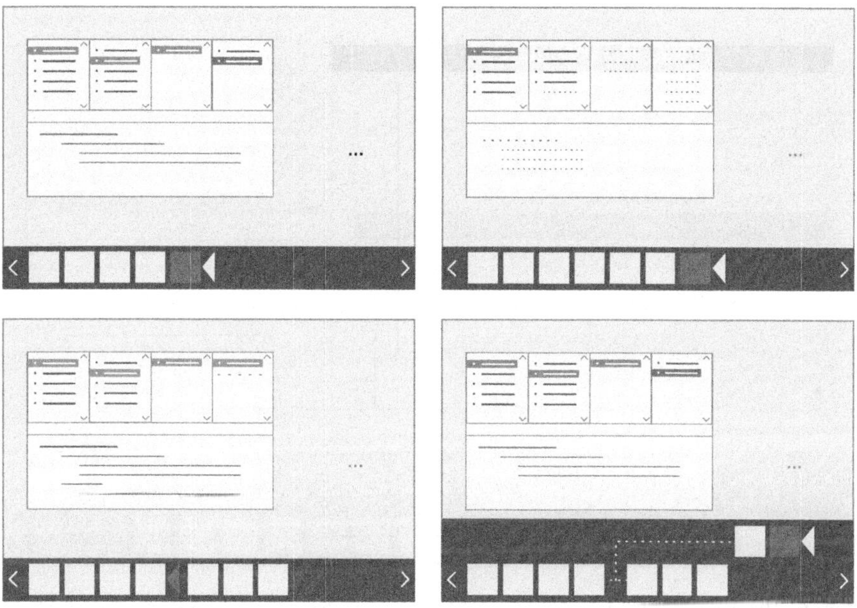

Fig. 4 (From *top left* to *top right*) A programmer modifies source code which implicitly creates items in the version bar. The programmer decides to withdraw several changes by going back to a previous version (*bottom left*), and continues working and creating new changes, which will appear on a new implicitly created branch (*bottom right*)

versioning and to allow for selecting previous versions, we have added a version bar (timeline) to the user interface of the programming environment (Fig. 3).

By continuously preserving intermediate development states, CoExist makes it easy for programmers to go back to a previous development state and to start over as shown in Fig. 4. Starting over from a former development state will implicitly create a new branch of versions. This preserves the changes the programmers want to withdraw, as they might be of use later on.

Fig. 5 Hovering shows which source code element has been changed (*left*). Holding shift in addition shows the full difference to the previous version (*right*)

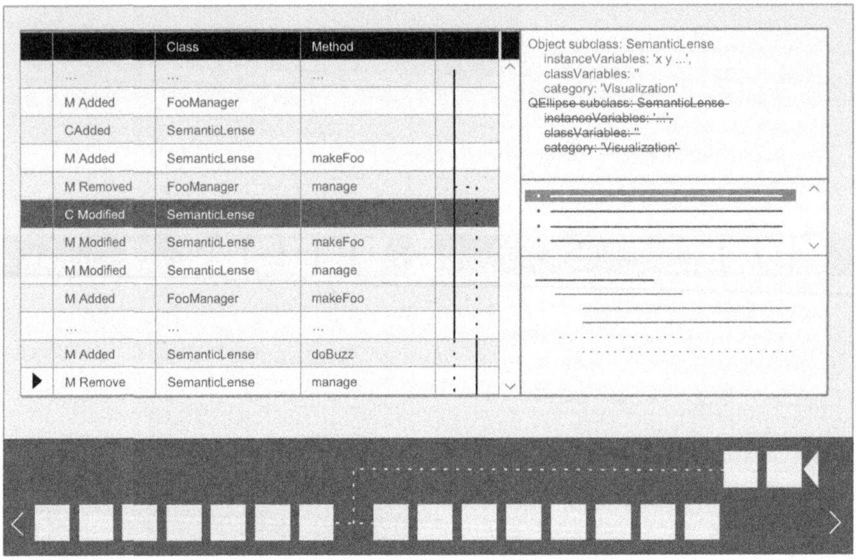

Fig. 6 The version browser provides a tabular view on change history. Selecting a row shows corresponding diff information in the panes to the right

CoExist provides two mechanisms dedicated to support programmers in identifying previous versions of interest. First, programmers can use the version bar, which will highlight version items that match to the currently selected source code element (Fig. 5). Hovering the items will display additional information such as the kind of modification, the affected elements, or the actual change performed.

Second, programmers can use the version browser to explore information of multiple versions at a glance. The version browser shown in Fig. 6 displays basic version information in a table view, which allows a fast scanning of the history for source code elements of interests.

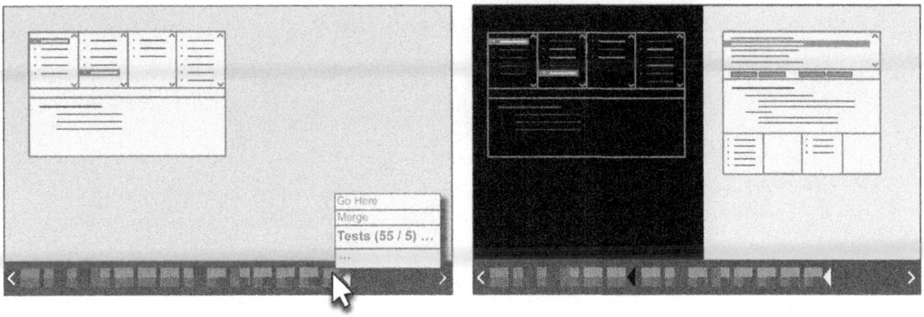

Fig. 7 The items in the version bar are now a visualization of the results of the tests that have been run in the background (*left*). A second inner environment allows the user to explore a previous version next to the current one (*right*)

The versioning facilities of CoExist also allows continuously running analysis on every newly created version. In particular, it supports running test cases to automatically assess the quality of the change made. The test result for a version is presented in the corresponding item of the version bar (Fig. 7). This makes the effect of each change regarding test quality visible. The user can also run other analysis such as performance measurements. CoExist provides full access to version objects and a programming interface to run code on them. Programmers can thus focus on their task at hand and, when necessary later on, can analyze the impact of each change.

When in the course of change programmers suddenly become curious about how certain parts of the source code looked previously or how certain effects were achieved, they can open a previous version in a separate working environment as shown on the right in Fig. 7. They can browse and explore the source code of a previous version and compare it to the current development version. In addition, they can also run and debug programs in this additional working environment.

With that, CoExist enables efficiently recovering knowledge from the previous version. This avoids the need for a precise understanding of every detail before making any changes.

With CoExist, programmers can change source code without worrying about the possibility of making an error because they can rely on dedicated tools that help with whatever their explorations will reveal. They no longer have to follow certain best practices to avoid undesired consequences of changing code.

5 Experiment Design

We hypothesize that programmers making use of CoExist will perform better in program design tasks that involve a strong degree of uncertainty. To gain empirical support for this claim, we have designed a controlled experiment. We have decided for a repeated measure factorial design, in which subjects are instructed to *improve* the source code design of given programs as best as possible in the time frame of 2 h.

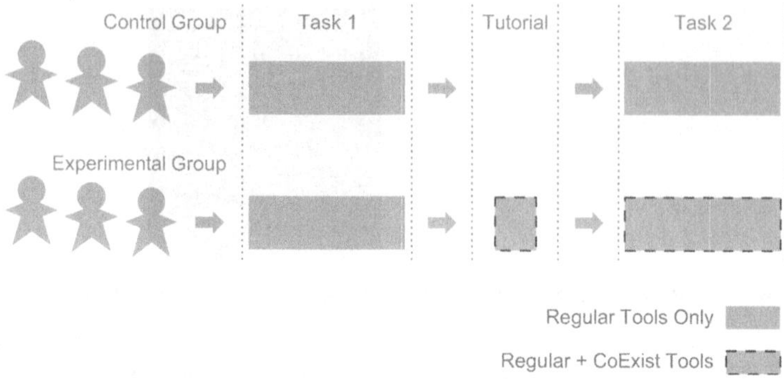

Fig. 8 Our experiment setup to compare performance in program design activities

5.1 Repeated Measure Factorial Design

Figure 8 illustrates the setup of our controlled experiment. Subjects are assigned to either of two conditions, the control group or the experimental group. Subjects in the control group use the regular development tools for both tasks. Subjects in the experimental group use regular development tools only for Task 1. After that, they receive a tutorial in CoExist and have time to try it out and experience its benefits and limitations. Thereafter, subjects in the experimental group work on Task 2 and can therefore make use of CoExist in addition to the regular tool support. So while for Task 1 all subjects use regular tools only, subjects in the experimental group may use CoExist for Task 2. After receiving task instructions and material, subjects had two hours for working on the respective task and achieving as much improvement as possible.

At the time of writing, 20 students have already participated in the experiment. They worked on both tasks on two different but subsequent days. Both tasks are scheduled for the same time of the day to ensure similar working conditions (hours past after waking up, hours already spent for work or studies, . . .). Typically, we scheduled the task assignments after lunch so that for day 2, there was time left to run the CoExist tutorial session upfront (before lunch time).

We try to keep subjects unaware of their assignment to the conditions, which worked well for day/task 1. However, on day 2, subjects in the experimental group could guess that they received special treatment because they were introduced to CoExist and were asked to make use of it. Nevertheless, subjects in the control group were unaware of the experimental treatment. They did not know that subjects of the experimental group run through a tutorial and can use CoExist for task 2. Hence, subjects were not entirely blind concerning the treatment, although we tried to be as close as possible to the blind setting.

This setup gives us two measures for every subject which will be prepared in tables such as shown in Fig. 9. Such a data collection allows tests for statistical differences between task 1 and task 2 as well as between the control and the

		Task 1	Task 2
Control Group	01		
	02		
	03		
	...		
Experimental Group	11		
	12		
	13		
	...		

Fig. 9 Data collection form corresponding to the experiment setup

experimental group. It also allows tests for the existence of an interaction effect of the two factors, which will indicate whether or not CoExist has an effect on programmers' performance.

5.2 *Task: "Improve"*

On each of the two days, subjects work on a different computer programs, but the task is the same. Participants are requested to improve the design of the source code. The two different programs are relatively small computer games like Tetris.

The procedure for each day is as follows. At the beginning of the assignment, subjects are introduced to the game. After introducing the game play subjects have a few minutes to play the game and get familiar with it. Afterwards subjects receive the assignment. The task is to study the source code, to detect design flaws in general and issues of unnecessary complexity in particular, and to improve the source code as much as possible in the given time frame of 2 h. To help understand the intent of the task, we provided sample descriptions of possible improvements, for example:

- Extract methods to shorten and simplify overly long and complicated methods
- Replace conditional branching by polymorphism
- Detect and remove unnecessary conditions or parameters

Subjects were told to imagine that they co-authored the code and are responsible for it, and that they now have time dedicated to improve the code in order to make future development tasks simpler (enhancements or maintenance).

For both tasks, the given program was a relatively simple single player computer games. Figure 10 shows screenshots of the game for Task 1 and Task 2. The game on the left is called LaserGame. The player's goal is to place mirrors in the field so

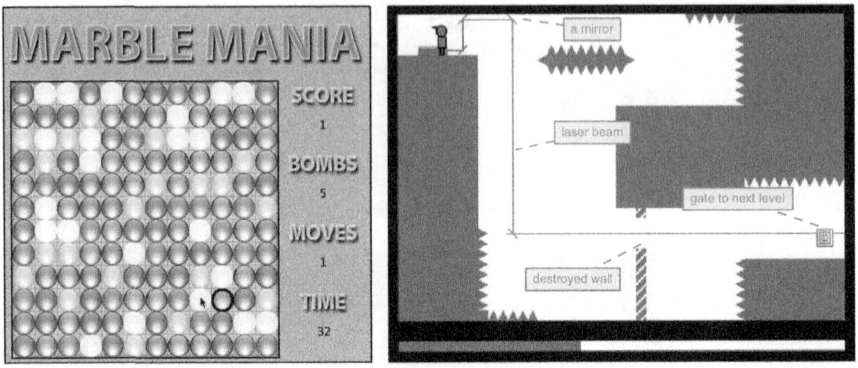

Fig. 10 Screenshots of the games used as experimental unit: MarbleMania (*left*), LaserGame (*right*)

that a laser is redirected properly to destroy the wall that blocks the way to the gate to the next level. The game on the right is called MarbleMania. The user has to switch neighbored marbles to create one or more sequences of at least 3 marbles that have the same color, either by row or by column.

Subjects are also asked to describe their improvements. They should imagine themselves as part of a development team. The description should help other team members to better understand their changes and how they improve the code.

We decided to use these small games as the experimental unit primarily due to practical reasons, but this choice turned out to be useful in unexpected ways. Such games are developed by students in one of our undergraduate courses. Over the years we have gained an interesting repertoire of game, all having different characteristics and qualities. Both of these games function properly and have a simple but still fun game play. So, only little time is required to get familiar with the program's functionality. Both games leave significant room for improvement in terms of the source code, as do most of the games developed in this course. Furthermore, both games come with a set of tests cases, which also have been developed by the respective students. Test cases are particularly useful when existing source code has to be changed because they are a means to automatically check whether selected aspects of the program still work as expected. However, while the offered test cases are useful, they are not sufficient. Manual testing of the games is necessary, as it is often the case in other projects.

Independent of these characteristics, using the games as the subject of work has another benefit. Participants of the study are primarily HPI students. And since they had to take our class, they all have gained similar experiences in developing such games. They have similar knowledge about the technologies, frameworks, and libraries used for developing these kind of games. Consequently, experience is not a factor that varies greatly among the participants.

5.3 Design Justification

For our experiment, we decided to keep the time factor fixed and measure the amount of achieved improvements. Keeping the time fixed is one of two typical ways to examine the effects of tools or methods. The other way is to fix the amount of work to get done by providing a clear unambiguous goal and measure the time needed to achieve this goal (Juristo and Moreno 2010).

A fixed time setup seems preferable for experiments that focus on design tasks, mainly because uncertainty is inherent to design tasks. The setup that measures time to completion requires a clearly defined task without any uncertainty or ambiguity. There has to be clear indicator when the task is finished. Also, the task description should provoke similar thoughts, so that all subjects have the same idea what the goal is and how it should be approached (given the defined conditions). These criteria can hardly be met in an experiment where participants should accomplish a design task.

Related research also shows that fixed time setup is a typical choice. In (Dow et al. 2009, 2010), for example, the authors report on the empirical examination of prototyping techniques. The response variable (dependent measure) has always been some form of quality criteria of the design outcome while participant have had a fixed amount of time to create the best possible design. However, we are unaware of an experiment report in the software engineering field that examines an effect in program design tasks.

Besides the decision for the fixed time setup, we have decided to rely on a repeated measurement experiment design in contrast to other options. A repeated measurement setup is preferable because programmers strongly vary in approaching such tasks. Programmers have different working speeds, which involves code comprehension, code writing (typing speed), but also tool usage. Furthermore, when programmers face the task to spend a fixed (and relatively short) amount of time on improving source code, some programmers will have the tendency to focus on the various small issues in the code, which are probably also easy to fix, while other programmers will have the tendency to try to identify major flaws in the overall program structure of the code and to improve on that while ignoring smaller issues. While there can be significant differences, different contributions can be similarly important for the long term success of a software project. However, the difference in the response variable between programmers can be large in relation to the possible difference caused by the provoked variations. For still being able to discover statistical effects in these circumstances, literature recommends repeated measurement experiment setups, in particular when having limited access to subject candidates (Juristo and Moreno 2010).

5.4 Analysis: Coding Changes

Besides a meaningful setup that allows for comparing performance, the experiment requires a meaningful quantitative response variable (dependent variable). Therefore, we have operationalized the notion that programmers can spend more or less time on contemplating source code and testing their ideas mentally before actually making the changes.

We assume that CoExist encourages participants to make changes as they think of them because they can always recover from undesired consequence easily. We also assume that without CoExist, participants will spend more time on thinking about their ideas before making changes to the code in order to avoid making errors.

In the context of the experiment, we believe that CoExist enables programmers to achieve more improvements in the given time frame. Every thought about a possible improvement, no matter how vague and uncertain it still may be, can either turn out appropriate or inappropriate. If the vague idea turns out to be appropriate, programmers who directly started making changes will clearly save time because they avoid spending time on upfront thinking. Findings in design research suggest that given a design task full of uncertainty, programmers who directly make the changes will save time in case that the idea turns out to be inappropriate. By making the changes, they support their thinking by doing and explore their ideas and their limits more efficiently. CoExist will help to recover fast and to get back to a desired development state.

While we assume that participants using CoExist will make more changes, the pure number of changes made, which correlates with number of versions, is not a meaningful proxy for the amount of achieved improvements. This is for various reasons:

- Changes that lead to new versions strongly vary in the amount of changed code and in the effect they have. While adding leading whitespace is a small change to the code, which likely has no effect on the program execution, removing statements or parameters from a method is a much larger change, which almost always will has a effect on the program execution.
- Participants might want to withdraw changes. In an extreme case, a subject might spend an hour of work on a particular idea to recognize later that the idea cannot work out properly. This will create many versions in the history which cannot be counted as improvements. Moreover, when not using CoExist, participants have to manually withdraw their made changes, which will in turn create additional changes.
- It might also be the case, that a series of changes was only good for inspiration and helped the programmer to develop a better idea of how to improve the elements of current interest. In this case, it would be unfair to double count the made changes, the changes made for the initial idea and also changes made for the final improvement.

Facing these constraints, a meaningful way to quantify the amount of achieved improvements is to code the changes that persist over time, which means to identify groups of related changes and assign them to categories. Such change categories can be either generic or rather specific to the given program.

Generic improvements are for example: to rename an instance variable; to replace a parameter with method; to make use of cascades; to inline a temporary expression; to replace magic string/number with method.

Improvements specific to the MarbleMania Game are, for example: to replace the dictionary holding "exchange state" with instance variables; to replace `isNil` checks in the destroyer with *null objects*; to remove button clicked event handling indirections.

To perform the coding, we have analyzed the data in two steps. In a first step, we have listed the timestamps of all versions (in a column of a spreadsheet) and separated them according to commits, with subjects made during the task. In the second column, we added the respective commit message (illustrated in Table 1). The commit messages help understand the intent of the changes, which gives the required context to understand the small changes to the individual source code elements. In a second step, we identified improvements that we assigned to a category. Thereby, a coded improvement can consist of only one actual change or it can involve many changes. Sometimes, all the changes made for one commit contribute to one coded improvement.

These change categories are assigned numbers that represent the relative effort required to implement them. We sum up these numbers for the coded improvements to finally get a measure of the amount of achieved improvements, which can be used to run the statistical tests.

6 Summary

Programmers spend time on source code design to better support current and future programming activities. While working on their code base, programmers can make errors. Making errors represents a risk because it often requires tedious and time-consuming recovery work. Traditionally, programmers have to anticipate such situations to keep the costs for recovery low. However, anticipating errors is not always possible, as we have illustrated by means of an experience report. We have also described that precautionary activities can easily be ignored and distract from the actual work.

We have presented CoExist as an extension to programming environments. CoExist has been designed to encourage change. It avoids the risk of error making by providing dedicated support to recover from undesired consequences. We believe that such a tool-based approach is preferable because it encourages programmers to support their thinking by doing. We have presented an experiment design to empirically examine this claim. We measure subject performance in two different tasks. In each task, participants have to improve the source code design of a given program. Their changes are recorded. Afterwards, the changes are coded to identify achieved improvements and to compute a quantitative measure of programmers' performance.

Table 1 Spreadsheet Excerpt with coded version data

Version timestamp	Commit message of participants	Coded improvements
...		
14:01:41	In LaserBeam extracted code that is similar in all these calculate methods; improved the previously extracted, generic calculateWay: ... method. (**two things happened** – refactoring + additional improvements) ... “Refactoring: summarized multiple similar methods into one, with 3 parameters. Original methods call the new one.”	LG_ExtractGenericCalculateMethod (ReplaceSimilarStatementsWithCallToExtractedMethod)
14:02:16		
14:02:32		
14:02:49		
14:03:01		
14:03:05		
14:03:11		
14:03:16		
14:06:08		
14:06:42		
14:06:46		
14:13:06	Simplified LaserBeam>>#setStartPoint, deleted useless condition, integrated code from called methods, and removed the other methods. “Simplified method based on detected ‘invariant’, that self laser direction is always 1@0 at this code location. Removed 2 methods.”	RemoveStatements + 2 * InlineMethod
14:14:18		
14:14:25		
14:15:16		
14:15:28		
14:15:38		
14:15:42		
14:17:25		
14:18:39	“Removed 2 unused variables of SWA18Laser. One seemingly was not used at all (point), the other (direction) got useless after previous commit. Removed all usage of the direction-variable (was only used in tests).”	2*RemoveStatement + 2*RemoveUnusedMethod + 2*RemoveUnusedInstVar
14:18:43		
14:18:53		
14:18:53		
14:18:53		
14:19:41		
14:20:33		
14:21:02		

14:25:02	"Extracted method in level parsing." #readNumberArrayFrom:	LG_LevelLoader_ExtractSimilarStatements
14:26:00		
14:26:08		
14:26:26		
14:26:53		
14:27:15		
14:27:20		
14:27:37		
14:27:50		TmpVarRenaming
14:29:20		TmpVarRenaming
14:29:29	"Removed useless method" – belongs to previous context	InlineMethod
14:30:18	"Removed another useless method" – namely #readTimeFrom: – integrated	Recategorization
14:30:59	the more meaningful one-liner in the caller	InlineMethod
14:31:11		
. . .		

References

Allen D (2001) Getting things done: the art of stress-free productivity. Penguin, New York
Apache Software Foundation (2009) Subversion best practices
Beck K (1996) Smalltalk best practice patterns. Prentice Hall
Beck K, Andres C (2004) Extreme programming explained: embrace change. Addison-Wesley Longman
Blackwell AF (2002) What is programming. In: 14th workshop of the Psychology of Programming Interest Group. Citeseer, pp 204–218
Dorst K, Cross N (2001) Creativity in the design process: co-evolution of problem- solution. Des Stud 22(5)
Dow SP, Heddleston K, Klemmer SR (2009) The efficacy of prototyping under time constraints. In: Conference on creativity and cognition
Dow SP, Glassco A, Kass J, Schwarz M, Schwartz DL, Klemmer SR (2010) Parallel prototyping leads to better design results, more divergence, and increased self-efficacy. ACM Trans Comput-Hum Interact (TOCHI) 17(4):18
Fowler M (1999) Refactoring: improving the design of existing code. Addison-Wesley Professional, Reading
Goldschmidt G (1991) The dialectics of sketching. Creativity Res J 4(2)
Juristo N, Moreno AM (2010) Basics of software engineering experimentation. Springer
Kirsh D (2010) Thinking with external representations. Ai Soc 25(4):441–454
Lim Y-K, Stolterman E, Tenenberg J (2008) The anatomy of prototypes: prototypes as filters, prototypes as manifestations of design ideas. ACM Trans Comput-Hum Interact (TOCHI) 15(2)
Lindberg CA (2008) Oxford American writer's thesaurus. Oxford University Press, New York
Parnas DL (1972) On the criteria to be used in decomposing systems into modules. Commun ACM 15(12):1053–1058
Steinert B, Cassou D, Hirschfeld R (2012) Coexist: overcoming aversion to change. In: Proceedings of the 8th symposium on dynamic languages, DLS'12, New York, ACM, pp 107–118
Suwa M, Tversky B (2002) External representations contribute to the dynamic construction of ideas. In: Diagrammatic representation and inference, vol 2317. Springer Berlin/Heidelberg

Design Thinking: Expectations from a Management Perspective

Holger Rhinow and Christoph Meinel

Abstract The authors present first results from an empirical study on the integration of Design Thinking in large corporations. In this article, we examine the managers' expectations towards Design Thinking in one large software-developing corporation. Empirical results from various interviews show that managers explicitly expect Design Thinking to contribute to existing frameworks such as Lean in general and Scrum in particular. It can be assumed, that the success of Design Thinking in this corporation will therefore depend on its compatibility with those established frameworks.

1 Introduction

Over the last decade, Design Thinking stepped out of the design course and received the global recognition of agencies and corporations. Corporate management in various industries currently drives the embedding of Design Thinking.

We investigate the relevance of Design Thinking from a management perspective. It is based on a long-term study at a multinational software vendor that embedded Design Thinking in its corporate structure. The study further compares different corporate initiatives and their impacts over a longer period.

In the following article, we focus on the management's expectations within a specific corporation. We therefore conducted interviews with managers at different levels of the organization, before they initiated Design Thinking projects, as well as after the projects ended.

H. Rhinow (✉) • C. Meinel
Hasso Plattner Institute at the University of Potsdam, Potsdam, Germany

H. Plattner et al. (eds.), *Design Thinking Research*, Understanding Innovation,
DOI 10.1007/978-3-319-01303-9_15,

2 Previous Research That Links to Our Long-Term Study

The growing interest in Design Thinking follows a changing discourse in research. The former descriptive design discourse became more and more normative and therefore compatible for corporations that are pursuing new ways to manage their processes and structures. Some previous research can be linked to the observation in our long-term study that management articulates specific expectations towards Design Thinking. Incidentally, the expectations that are mentioned can be found in other corporations as well.

Design Thinking has been associated with various descriptive and normative discourses in design research over the last 50 years (Plattner et al. 2009).

Peter Rowe (1987) initially coined the term Design Thinking back in its infancy during an ongoing design discourse in the 1960s. The discourse, among other things, dealt with the nature of design as a process and was influenced in multifaceted ways by theories and observations in cybernetics (Simon 1999), social planning (Rittel and Webber 1973), and practical design (Schoen 1984). The discourse was dominantly descriptive, providing insights into the work of scientists, social planners, architects, designers and other reflective practitioners (Schoen 1984) who were dealing with complex conditions, namely recursive systems (von Foerster 2002) or wicked problems (Buchanan 1992; Rittel and Webber 1973).

Over the last 20 years the discourse has became dominantly normative, describing Design Thinking as a yet-to-be-achieved-condition, among other things: a not-yet-developed *innovative working culture* of creative teams and networks (Brown and Watt 2010), a not-yet-captured chance of *creating user-focused solutions* (Badke-Schaub et al. 2010), a not-yet-achieved balance between *learning and knowing* (Bucolo and Matthews 2010), a not-yet-lived but highly promising process of switching between *spaces* (Brown 2008), a not-yet-achieved reconciliation of *analytical and intuitive thinking* (Martin 2009) or a not-yet-empowered workforce of *non-professional-but-could-be-designers* (Dunne and Martin 2006).

The normative discourse is gaining significant attention in the business world, as large banks, software vendors, and multinational manufacturers set up initiatives to embed Design Thinking in their development and manufacturing processes (Clark and Smith 2008). First, corporations develop initiatives to raise awareness for Design Thinking, and they then commit themselves to empower their own employees by setting up education programs and Design Thinking projects (Carlgren et al. 2011).

3 Case Study and Research Approach

As a starting point for our research we present the observation that management in various corporations embed Design Thinking with specific intentions to improve the status quo.

The following case study focuses on the global embedding of Design Thinking at a multinational software vendor.

The company is a large, multinational software vendor. It is a DAX corporation that became highly successful over four decades by providing standardized business software solutions for a wide range of industries.

In our research project for this case study, we conducted over 50 semi-standardized interviews with managers, team members and Design Thinking coaches that were all part of a large Design Thinking initiative consisting of more than 40 projects over a period of 6 months. Most of the projects started in the spring of 2012 and ended in the fall of 2012. A majority of projects in the initiative were situated in Europe, others were located in Asia, the US and Canada.

The projects covered a range of innovation topics, such as new product development, process improvement, reorganization, and the evaluation of innovative basic technologies for the software industry.

We investigated 11 different projects, and conducted interviews with all responsible project managers, the so-called product owners, with line managers, a vice president and at least one team member and a coach of each project. The interviewed product owners, line managers, and the vice president had not taken part in previous Design Thinking projects, apart from smaller learning workshops. We were particularly interested in understanding their expectations of Design Thinking before they gained experience in this field. After the project, we conducted further interviews in order to investigate how their previous expectations matched the actual course of their projects and the repercussions of this. The interviews lasted from 45 to 120 min. During our research, we constantly reviewed and updated our questionnaires whenever new relevant observations indicated new topics of interest. This qualitative and iterative process is compliant with the rules of the Grounded Theory approach (Glaser and Strauss 2009).

4 Previous History to the Design Thinking Initiative

With regard to the case study, Design Thinking did not just appear but became a topic of interest in the last decade. During that time, the corporation was undergoing major changes. In our interviews, stakeholders mentioned several streams of events that had impacted the corporation significantly. Due to space constraints, we focus on three aspects that seemed to highly influence managers' expectations towards Design Thinking.

4.1 The Founder's Vision

One of the founders of the corporation is still a member of the advisory board and widely regarded as a very influential person within the corporation. The founder became an active supporter of a more design-oriented approach, influenced by the successful Design Thinking agency IDEO. Based on the perception of several interviewees, the founder actively positioned Design Thinking as a priority in the top management.

4.2 The Introduction of Lean

Three years before the initiative took place, the corporation started to undergo a major reorganization that is still being perpetuated today. In order to become a Lean organization, the corporation changed not only processes and structures but also introduced completely new positions, especially in development units, that make up a major part of the workforce.

Lean includes a variety of frameworks that often empower teamwork for different purposes, e.g. agile frameworks in the software development processes (Kittlaus 2012). Teams play a major role in this corporation, similar to other modern corporations. Previous differentiations and specializations have resulted in highly distinct work units with a perceived lack of knowledge transfer and collaboration. While these units led to higher degrees of efficiency, it was also apparent that corporate structures had become less flexible and agile in order to adapt to changing markets. As a consequence, corporations began to install different team-based approaches to establish a more creative working culture:

> The Lean approach brings back the strengths of the company in its beginnings, when the number of employees was a two or three digit number. (translated from Mackert et al. 2011, p. 48)

A major goal anticipated with the incorporation of Lean is the banishment of all sorts of wasteful activities and processes with the end result of becoming a more efficient corporation. Lean activates the creative potential of employees to this end, based on modern organization and personnel management (Stürzl 1996).

4.3 The Introduction of Scrum

Strictly speaking, Scrum is one Lean framework among others. Nevertheless, with regard to the case study, Scrum is of high importance. Scrum is a framework to empower teamwork in order to efficiently manage software development processes.

The incorporation of Scrum puts a strong emphasis on cross-functional teamwork and shared responsibilities during the software development process:

> Split your organization into small, cross-functional, self-organizing teams. Split your work into a list of small, concrete deliverables. Sort the list by priority and estimate the relative effort of each item. (Kniberg and Skarin 2010, p. 3)

4.4 Empirical Results from Our Case Study

Our research indicates that managers' expectations stem from perceived shortcomings in the corporation. All interviewed managers consensually linked their expectations of Design Thinking with needs to improve specific aspects in their corporation. While some expectations seem to reflect unique issues from units, teams or markets, other expectations were shared among most, if not all of the managers.

Managers consensually see a need to further improve the teamwork at all levels of the corporation. They expect to continue to work with established team-based approaches from Lean, such as the Scrum framework. However, they see shortcomings with regards to the existing frameworks. They directly link their expectations towards Design Thinking with regard to these shortcomings.

In the following, we highlight two of the major expectations we have observed in our case study.

4.5 Managers Expect Teams to Deliver User Value Propositions

With regard to the principles of Lean, managers expect their employees to challenge existing thoughts on user values and needs. Strictly speaking, managers expect their employees to deliver user-valid innovations that are worth pursuing in the light of limited resources:

> For me, from my understanding, Design Thinking is a concrete tool to support innovation and creativity. First of all, to understand what we should actually build for our customers. And the Lean principles that we have, in my opinion, focus on how we can get what we want to build, in a reasonable quality with a motivated team. (Interview with Product Owner#2 March 2012)

In order to understand why managers are particularly interested in user value propositions for further developments, it is important to reflect on Lean. Lean is perceived as a management framework that consists of multiple frameworks to empower teamwork within the corporation. It is not only restricted to development processes but became a potential advanced and complex framework for almost all organizational structures and processes (Stürzl 1996).

> Lean thinking is much more than tools such as kanban, visual management or queue management or merely the elimination of waste. As can be seen at Toyota, it is an enterprise system resting on the foundation of managers-teachers in lean thinking with the pillars of respect for people and continuous improvement. Its successful introduction will take years and requires widespread education and coaching. (Larman and Vodde 2009, p. 90)

Or, as one interviewee put it:

> Can anything that happens within the corporation be seen through the lenses of Lean? I think that is what we want to achieve. (Interview with Product Owner#2 March 2012)

Japanese corporations and their early adopters in the US and Europe (e.g. Porsche) achieved efficiency gains by the integration of lean that became a global topic of interest (Womack et al. 1997). The empowered teams become the major driver to constantly improve processes and structures.

Lean as a management framework does not imply a concrete catalogue of decisions to be made but rather implies generic guidelines. These support teams in coming up with decisions themselves on how to improve their specific situation in the corporation. In a nutshell, those teams are, while they are on the job, empowered to sense suboptimal conditions by distinguishing if existing structures and processes are of value or of waste for the corporation. In other words, Lean corporations are conditioned to minimize all factors that are not creating value (Stürzl 1996).

Value is defined with regard to a user, the customer and her needs. It includes all actions that directly and indirectly contribute to the development of a proposition for the user, in this case the customer:

> Value - the moments of action or thought creating the product that the customer is willing to pay for. In other words, value is defined in the eyes of the external customer. (Larman and Vodde 2009, p. 58)

One of the managers summarized his understanding of value:

> At the end of the day, we can only sell software that has added value for our customer. It is of value, if the customer is of value for us. (Interview with Product Owner#8 March 2012)

Any actions that do not comprehensibly contribute to the value proposition are suspicious to be waste and may become a condition that should be banished. This can include small actions such as holding unproductive meetings or unnecessary delays to formal routines such as complete development processes that deliver outdated and therefore useless outcomes:

> Waste - all other moments or actions that do not add value but consume resources. Wastes come from overburdened workers, bottlenecks, waiting, handoff, wishful thinking, and information scatter, among many others. (Larman and Vodde 2009, p. 58)

Managers and employees perceive the given framework of Lean as highly useful but not yet sufficient:

> For me (...) Lean is focusing very strongly on efficiency. (Interview with Product Owner#2 March 2012)

The important prerequisite for teams is therefore to have a clear understanding of a value proposition; otherwise it is impossible to determine which actions contribute value and which do not. The managers acknowledge the need for a clear value proposition that their target users, e.g. customers, are willing to pay for:

> If target users tell you 'we want this and that', and their boss tells you 'I will not pay extra, that is not on the list, I don't need that'. It is a difficult situation. (...) You really have to

> consider the value that people are actually willing to pay for. (Interview with Product Owner#8 March 2012)

It is not trivial to find out clear value propositions, because e.g. different customers provide different perspectives:

> You have a customer contact people from development. We often deal with administrators on the other hand. The customer value is quite ambiguous. (Interview with Product Owner#8 March 2012)

From a management perspective, the definition of a value proposition – if not multiple value propositions – is not sufficiently covered by the principles and guidelines that mark the Lean framework. The need for a clear understanding of the customers' needs becomes a top priority for this corporation, however communication with customers was already challenging in the past:

> With Lean, one said, it is about the customer and participation, but I know many, many projects, that have never seen a customer. It is tough finding customers, it is not that we cannot find them, quite the opposite, but few of them are willing to invest their time. Or they simply don't understand what we actually want. (Interview with Product Owner#4 May 2012)

Not only does management perceive a lack of workable value propositions, it also acknowledges the issue that value propositions may change constantly while needs and market requirements change. Therefore, a value proposition for a customer that may have been valid in the past is not necessarily valid in the present or near future. They need to adapt to instable markets, like organic systems. Employees are therefore responsible for finding out the corporation's challenge (Baecker 2007).

The lack of a process to define value propositions had been perceived before Lean was introduced. It led to reactions such as the establishment of user research in development projects, and a growing amount of market research.

In the managers' perception, Design Thinking is a new alternative to rethink the definition of value proposition for the corporation's customers:

> Yes, I think (our corporation) established Lean two years ago in the context of agile development. Teams of ten were formed with new roles away from the pure line organization to a more agile organization. The question was answered as to how to carry out development. Now, from my understanding, Design Thinking is a possibility to define what to develop. (Interview with Product Owner#3 April 2012)

Design Thinking in this understanding also works as a framework to empower teamwork, giving guidelines on how to come up with decisions with regard to value propositions.

In our case study, the Design Thinking framework is a project-based approach to follow a process of different, iterative steps that empower teams to define and illustrate values of clearly defined users, such as customers. Teams work in this framework by approaching a very distinct potential issue or user's problem, represented by a constructed, hypothetical user, a so-called persona (Cooper 2004). A user-centered starting point is defined, e.g. instead of a population-based survey as it is often done in market research. During the process the team redesigns the initial project briefing as a starting point for their design challenge; they iterate on the nature of the problem, while they approach different possible solutions, they focus on specific user issues and synthesize them, they ideate upon self-constructed questions, and they prototype their ideas in a way they perceive as usable for further testing and iterations.

The teams intuitively navigate through a stream of options and make numerous decisions even though they might not explicitly reflect their actions as decision-making. In fact, every meaning that is abstracted from user observations, every idea that is developed and linked to a cluster of prior ideas, and every design detail that is built as part of a prototype is in fact a design decision that is made along the process (Schoen 1984). The managers expect teams to work and decide autonomously to a certain degree, similar to Lean:

> I mean, we work in a Lean mode and our employees who work together on a topic organize it themselves. It is not the case that everyone is doing stuff alone. (Interview with Product Owner#7 April 2012)

The framework therefore gives room for creation, testing and iterating but it also restricts the team to deliver a value proposition that is illustrated by a prototype. The team, for example, cannot decide on their own to directly develop new software solutions.

It is assumed that those tangible definitions of value can be used to distinguish between value and waste. In this sense the results of Design Thinking can be used as a prerequisite for decision-making in the Lean framework.

4.6 *Managers Expect Teams to Deliver Operability for Software Development Projects*

During the last 3 years, Scrum became the default framework to develop complex software development processes.

Scrum as a specific framework is in an exposed position within the holistic framework of Lean. It is incorporated in order to empower specific teams, so-called teams of ten, to manage themselves to a certain degree in order to efficiently develop software solutions.

Teams of ten consist of a product owner who is responsible for the solution and the prerequisites, a crossfunctional team that designs, codes, tests, documents, and a scrum master who facilitates the process (Grau 2012). They work in so-called sprints in which they commit themselves to developing certain objects of the final software solution. The teams usually meet every day for a short sprint review to reflect on the prior tasks and define what tasks need to be done that day.

Within the corporation, Scrum is perceived as a powerful framework to facilitate work by responsible team members independently, as they know well how much work they can do in a certain time. The team members do not need to wait for managers to provide orders. The sequence followed in Scrum is iterative to a certain degree as mandatory customer reviews can lead to certain changes in the development process, even though managers doubt the iterative aspect:

> (Normally) in these teams of ten, (. . .) you have this feeling that you are doomed to success. (Interview with Product Owner#8 March 2012)

The centerpiece of every Scrum project, and its most important artifact, is the so-called product backlog. The product owner is responsible to initially develop a backlog, a collection of prioritized backlog items that are distinct work tasks, which in total define a complete software solution. The later distribution of tasks is done by the team members themselves, leading to a higher self-organization.

Depending on the accuracy of the backlog estimation, the final allocation of all executed backlog items will inevitably lead to a valid and functional software solution, no matter which developer is responsible for certain tasks and no matter in which sequence they are executed.

The backlog illustrates a complete overview of all tasks that are needed to be done in order to come up with the software solution. As we have seen with regard to the overall Lean framework, every action, including all software developments, needs to be based on a value proposition and therefore is legitimately creating value for the corporation. This means that the initial backlog must be based on a value proposition. Equally important, the backlog needs to address the value proposition adequately, meaning that the backlog items are a valid composition in order to deliver an evident-based, user-desired result.

It is important to realize that a value proposition itself is not a user-desired result. The value proposition is synonymous with a statement of what a user desires and how much she is willing to pay for it. A user-desired result is a statement of how to address the desire with regard to a product or service solution. While a value proposition may be ambiguous, the solution is always a concrete statement.

The backlog therefore represents a product owners' interpretation of a concrete solution. During the Scrum process the interpretation may change because customer reviews indicate the necessity to do so, but it will always be expressed in a concrete statement.

At this point, managers perceive a shortcoming with regard to Scrum. The backlog only represents a concrete solution, if someone has done the transfer from a rather ambiguous and complex value proposition to a concrete solution statement. As previous design research shows, user problems and desires can be understood as complex, so-called wicked problems (Rittel and Webber 1973). Keeping with this terminology, a backlog can only represent so-called tame problems, e.g. user solutions that can be calculated, e.g. broken down into smaller parts:

> 'Tamed problems,' such as mathematical problems are causal microworlds that we, through enormously inventive interpretive skills are at times able to use to ensure that trains run on time, computers calculate bank balances, and bridges do not sway out of line. (Coyne 2005, p. 9)

There needs to be a prior contribution, in which those wicked problems are somehow processed or redesigned into tame problems. In this case it is the sum of distinct tasks that when combined result in a concrete and also complete solution statement:

> It is an important aspect (. . .) at some point you need to focus on things, so to speak out of the cloud, and to say now comes the transfer to development. (Interview with Product Owner#5 April 2012)

According to Glanville the processing of complexity into something simple is one of the characteristics that define designers:

> Designers create, from a situation presented as complex, something simple: a unique, new whole, through the making of which they handle and reduce complexity. (Glanville 2011, p. 33)

Product owners, who own the product backlog, are responsible for addressing this issue, and they therefore disassemble the overall challenge, or the overall goal:

> First of all, I mean, we have a certain period of time to develop, we also have to consider to disassemble and develop our insights that we identified while visiting our customers. This is our operating environment, actually. (Interview with Product Owner#7 April 2012.

This approach leads to new challenges:

> But also the product owner, Scrum, has huge problems to disassemble problems into subproblems, which he normally discusses with his architect or the whole team. It is a common problem, if he disassembles problems in subproblems, people tend to stop thinking while working on the problem. I call it this a working mode of cognitive relief. If a problem is disassembled, I simply don't understand the context anymore and I tell myself to just do it. (Interview with Scrum Master September 2012)

The managers, especially the product owners, expect Design Thinking to be a framework that takes into account the need to disassemble complex problems. They expect concrete solution statements. Design Thinking is therefore positioned as a prerequisite in order to efficiently work in Scrum:

> I suppose that we will get our input for backlog items out of the method of Design Thinking itself. I mean, what we will continue to do is the planning for our sprints, and there we will integrate these backlog items out of Design Thinking. (Interview with Product Owner#7 April 2012)

Especially product owners are keen to share responsibilities with teams in order to develop operable results out of user insights:

> It is not the classic situation anymore, where a product owner says: 'I am responsible, I make the decisions'. I believe, we, as a highly motivated team, will try to see, what we can get out of it, this is my approach. Of course, at some point you have to make a decision, but I believe it is the right approach to see what insights our people get from our customers and to give everyone a voice in order to decide what we are going to do with it in the end. (Interview with Product Owner#3 April 2012)

Managers seem to expect Design Thinking to happen before teams switch into the Lean framework:

> Okay, I will do Design Thinking to find out, what I can do for the end-user. And (then) I develop a real product. That would be a consequence (...) effectively developing an outcome after Design Thinking, this is part of Lean. This is suitable for me. (Interview with Product Owner#6 May 2012)

The backlog as an interface will answer the compatibility between Design Thinking and Scrum:

> And this is fixed. You can provide best practices. You can say Design Thinking illustrates tasks in a certain way, as a tool. (...) We need to come together and to consider how we can integrate Design Thinking into the corset given by Scrum. This is the question: can we represent a backlog via Design Thinking, or can't we? (Interview with Product Owner#8 March 2012.

The expectations were nurtured through the Design Thinking discourse that described Design Thinking as a form of abductive reasoning to deal with wicked problems, while other frameworks such as Scrum can be characterized as deductive reasoning (Martin 2009):

> The linear model of design thinking is based on determinate problems which have definite conditions. The designer's task is to identify those conditions precisely and then calculate a solution. In contrast, the wicked-problems approach suggests that there is a fundamental indeterminacy in all but the most trivial design problems -problems where, as Rittel suggests, the 'wickedness' has already been taken out to yield determinate or analytic problems. (Buchanan 1992, p. 36)

Within the above-mentioned Design Thinking framework, teams are empowered to deliver tangible, concrete statements with regard to potential value propositions. They become concrete by visualizing and prototyping their ideas (Schrage 1999).

From a management perspective, the prototype may work as a valid and concrete solution statement but does not necessarily indicate a complete solution. In order to derive to a backlog that represents a complete solution statement, managers at this corporation embedded further artefacts that derive from Design Thinking and can be directly used to develop a backlog.

Especially two artefacts stand out in the perception of the managers: the so-called business model canvas (Osterwalder and Pigneur 2010) and the so-called user story map (Hildenbrand and Meyer 2012).

The business model canvas forces the team to define the value proposition of their prototype explicitly and with regard to the business environment it is supposed to fit in, e.g. the key customers that are addressed, the key partners it needs to deliver the value proposition, the estimated cost structure.

The user story map forces the team to define the value proposition as a sequential process from a user's perspective. User Stories may be more abstract but can be operationalized into working tasks with regard to a backlog:

> There is a tool, it is called user story mapping. It comes from Lean and it is used to create backlog items out of high level stories. (Interview with General Manager December 2011)

Managers expect to work on user story maps after their idea is developed and before they develop the solution:

> I do user story maps once I am a bit further in my process, that is what I would say. Design Thinking is more at the front end, when I want to develop my idea. (Interview with Product Owner#5 March 2012)

5 Summary

Managers clearly stated the expectation that Design Thinking contribute in a complementary fashion to existing frameworks, such as the holistic framework of Lean and the development-specific framework of Scrum.

Managers perceive Design Thinking as a powerful, yet limited framework to empower teamwork in order to develop promising value propositions that continuously improve processes and structures according to Lean. Furthermore and with regard to development projects, managers expect Design Thinking teams to operationalize these value propositions into concrete solutions statements, including a visual representation (prototype), a defined business context (business model canvas), and a complete definition of all activities that are related to the solution (user story map).

6 Outlook

Even though our research already indicates a change of the corporate culture beyond innovation projects, managers do not perceive Design Thinking to become a trigger to change the overall working culture:

> I see Design Thinking as a potential complement. It depends on what you want to develop. I cannot see us starting a Design Thinking project for everything we do. It is useful, whenever we explore open territory. (Interview with Product Owner#5 April 2012)

Beyond that, our research indicates that management will face new challenges and opportunities if Design Thinking becomes a consistent part of the corporation. Since teams are empowered to work independently and make design decisions on their own, to a certain degree, management will need to position itself at the interface between Design Thinking and Lean, Design Thinking and Scrum, as well as other frameworks that are compatible with Design Thinking but are not discussed in this article.

Since Design Thinking Teams may come up with a lot of potential value propositions that may even contradict themselves, it needs a management that makes decisions in terms of what value propositions may be useful as prerequisites for Lean initiatives and what definitions may be inconsistent with existing value propositions. Management in this context may act as an enabler for decentralized teams to either transfer or withhold value propositions into other parts of the corporations, with regard to consistency with overall strategies.

Management will also need to position itself at the interface between Design Thinking and Scrum in order to couple or decouple solutions statements from the later development process.

Also, it is still undecided how customer reviews in Scrum may lead to reconsiderations of the validity of solutions statements coming out of Design Thinking and, as a consequence, may demand a major change, e.g. by returning to the framework of Design Thinking. It is assumed that this will be a highly probable scenario in many development processes. However, it could also lead to new challenges for management of how to deal with the contingency of iterative frameworks in a business world of more or less fixed development cycles and delivery dates.

Our research will further investigate how managers approach these upcoming challenges and whether Design Thinking will match their expectations in the long run.

Due to legal standards, the authors currently withhold the name of this specific company and of the cited persons.

References

Badke-Schaub P, Roozenburg N, Cardoso C (eds) (2010) Design thinking: a paradigm on its way from dilution to meaninglessness? Design thinking research symposium, p 8

Baecker D (2007) Studien zur nächsten Gesellschaft. Suhrkamp

Brown T (2008) Design thinking. Harvard Business Rev, pp 84–92

Brown T, Watt J (2010) Design thinking for social innovation. Stanford Soc Innov Rev 8(1):28–35

Buchanan R (1992) Wicked problems in design thinking. Des Issues 8(2):5–21

Bucolo S, Matthews J (eds) (2010) Using a design led disruptive innovation approach to develop new services. In: Kobe C, Goller I (eds) Proceedings of the 11th international cinet conference: practicing innovation in the times of discontinuity, CINet

Carlgren L, Elmquist M, Rauth I (2011) Implementing design thinking – an exploratory study of large companies using design thinking in innovation efforts. In: Proceedings of the 2011 Tsinghua-DMI international design management symposium HK in Hong Kong, pp 1–19

Clark K, Smith R (2008) Unleashing the power of design thinking. Des Manag Rev 19(3):7–15

Cooper A (2004) The inmates are running the asylum. Sams, Indianapolis

Coyne R (2005) Wicked problems revisited. Des Stud 26(1):5–17

Dunne D, Martin R (2006) Design thinking and how it will change management education. An interview and discussion. Acad Manage Learn Educ 5(4):512–523

Glanville R (2011) Designing complexity. Revue für Postheroisches Management 8:24–41

Glaser B, Strauss A (2009) The discovery of grounded theory: strategies for qualitative research. Transaction Publishers, New Brunswick/London

Grau R (2012) Requirements engineering in agile software development. In: Software for people. Springer-Verlag, Berlin/Heidelberg, pp 97–120

Hildenbrand T, Meyer J (2012) Intertwining lean and design thinking: software product development from empathy to shipment. In: Software for people. Springer-Verlag, Berlin/Heidelberg, pp 217–237

Kittlaus H-B (2012) Software product management and agile software development: conflicts and solutions. In: Software for people. Springer-Verlag, Berlin/Heidelberg, pp 83–96

Kniberg H, Skarin M (2010) Kanban and scrum. Making the most of both. C4 Media, USA

Larman C, Vodde B (2009) Scaling lean & agile development. Thinking and organizational tools for large-scale Scrum. Addison-Wesley, Upper Saddle River

Mackert O, Hildenbrand T, Podbicanin A (2011) Von wasserfallartigen Softwareentwicklungsmodellen hin zu "Lean software product development". *Gesellschaft für Informatik e. V., Fachausschuß Management der Anwendungsentwicklung und -wartung (WI-MAW) im FB Wirtschaftsinformatik,* pp 48–51

Martin RL (2009) The design of business: why design thinking is the next competitive advantage. Harvard Business Press

Osterwalder A, Pigneur Y (2010) Business model generation: a handbook for visionaries, game changers, and challengers. Wiley, Amsterdam (self-published)

Plattner H, Meinel C, Weinberg U (2009) Design thinking. mi-Wirtschaftsbuch, Munich

Rhinow H (2011) Translated interview with general manager on the 14th of December 2011

Rhinow H (2012a) Translated interview with product owner#2 on the 23 March 2012

Rhinow H (2012b) Translated interview with product owner#3 on the 13 April 2012

Rhinow H (2012c) Translated interview with product owner#4 on the 14 May 2012

Rhinow H (2012d) Translated interviews with product owner#5 on the 22 March, the 13 April, and the 30 May 2012

Rhinow H (2012e) Translated interview with product owner#6 on the 14 May 2012

Rhinow H (2012f) Translated interview with product owner#7 on the 13 April 2012

Rhinow H (2012g) Translated interview with product owner#8 on the 23 March 2012

Rhinow H (2012h) Translated interview with scrum master on the 13 Sept 2012.

Rittel H, Webber M (1973) Dilemmas in a general theory of planning. Policy Sci 4(2):155–169

Rowe P (1987) Design thinking. MIT Press, Cambridge, MA

Schoen D (1984) The reflective practitioner: how professionals think in action. Basic Books, New York

Schrage M (1999) Serious play. How the world's best companies simulate to innovate. Harvard Business School Press, Boston

Simon H (1999) The sciences of the artificial. MIT Press, Cambridge, MA

Stürzl W (1996) Business reengineering in der Praxis. Durch umfassende Veränderung im Unternehmen einen Spitzenplatz im Wettbewerb erreichen. Junfermann, Paderborn

Von Foerster H (2002) Understanding understaning: essays on cybernetics and cognition. Springer-Verlag, New York

Womack J, Jones D, Stotko E (1997) Auf dem Weg zum perfekten Unternehmen. Wilhelm Heyne Verlag, Munich

The manufacturer's authorised representative in the EU is Springer Nature Customer Service Centre GmbH, Europaplatz 3, 69115 Heidelberg, Germany. If you have any concerns regarding our products, please contact ProductSafety@springernature.com

Printed and bound by CPI Group (UK) Ltd, Croydon, CR0 4YY
22/09/2026
02221064-0011